Praise for *God & t*

"Truly inspiring.... A lyrical, meditati
between science and Kabbalah. Prov. ountless insights—small and large—about God, consciousness, Torah and living in the universe.... I have rarely felt so cared for and so moved by an author/teacher."

—**Rabbi David W. Nelson**, author, *Judaism, Physics and God: Searching for Sacred Metaphors in a Post-Einstein World*

"Admirable clarity and wit.... Will be appreciated by everyone who takes an inclusive approach to the riddles of creation and the creator."

—**Timothy Ferris**, author, *Coming of Age in the Milky Way*

"Incredibly rich and immensely practical.... Anyone interested in science and spirituality will find this account of interest."

—**ALA** ***Booklist***

"Matt's timely revision of his classic book accurately recounts key recent surprising cosmological discoveries while his mastery of the *Zohar* helps him reveal and interweave their important spiritual messages."

—**Howard A. Smith, PhD**, Harvard-Smithsonian Center for Astrophysics; author, *Let There Be Light: Modern Cosmology and Kabbalah, a New Conversation between Science and Religion*

"Daniel Matt is this generation's greatest poet/scholar of Kabbalah. In a thoroughly contemporary way, he weaves the wisdom of ancient mysticism with the honesty and questioning of science to produce a rare treasure: a book that will enlighten, transform, elevate and educate!"

—**Rabbi Bradley Shavit Artson, DHL**, author, *God of Becoming and Relationship: The Dynamic Nature of Process Theology* and *Renewing the Process of Creation: A Jewish Integration of Science and Spirit*

"Tackles profound, difficult and complex issues and makes them understandable."

—**Rabbi Laura Geller**, Temple Emanuel, Beverly Hills, California

"Contributes in a significant and distinctive way to the contemporary dialogue between science and theology."

—**Robert J. Russell, PhD**, Christian theologian; director, Center for Theology and the Natural Sciences

"Playful, imaginative and thoughtful ... carries the dialogue between religion and science to new levels."

—***Publishers Weekly***

"A lively and fascinating volume that should provide the basis for much further dialogue and discussion."

—**Dr. Arthur Green**, author, *Ehyeh: A Kabbalah for Tomorrow*

"Revel in the informed, unexpected correlations this master of our mystic literature finds between his Kabbalah and our cosmologists' learned imaginings."

—**Rabbi Eugene B. Borowitz**, *Sh'ma*

"This poetic book ... helps us to understand the human meaning of creation."

—**Joel Primack**, leading cosmologist; professor of physics, University of California, Santa Cruz

"Highly recommended for the general reader."

—***Library Journal***

GOD & THE BIG BANG

2ND EDITION

Discovering Harmony between Science & Spirituality

Daniel C. Matt

For People of All Faiths, All Backgrounds

JEWISH LIGHTS Publishing

Nashville, Tennessee

God & the Big Bang, 2nd Edition:
Discovering Harmony between Science & Spirituality

Library of Congress Cataloging-in-Publication Data
Names: Matt, Daniel Chanan, author.
Title: God & the big bang : discovering harmony between science & spirituality / Daniel C. Matt.
Other titles: God and the big bang
Description: 2nd edition. | Woodstock, Vermont : Jewish Lights Publishing, [2016] | ©2016 | Includes bibliographical references and index.
Identifiers: LCCN 2015050761 | ISBN 9781580238366 (pbk.) | ISBN 9781580238472 (ebook) | ISBN 9781683360803 (hc)
Subjects: LCSH: Spiritual life—Judaism. | Big bang theory. | Jewish cosmology. | God (Judaism)
Classification: LCC BM723 .M384 2016 | DDC 296.3/75—dc23 LC record available at http://lccn.loc.gov/2015050761

Manufactured in the United States of America
Cover design: Tim Holtz
Cover art: Star Cluster NGC 2074 image from the NASA Hubble telescope.
Interior design: Glenn Suokko, modified by Tim Holtz

For People of All Faiths, All Backgrounds
Published by Jewish Lights Publishing
An Imprint of Turner Publishing Company
4507 Charlotte Avenue, Suite 100
Nashville, TN 37209
Tel: (615) 255-2665
www.jewishlights.com

To our daughter, Michaella,
and our son, Gavriel, who
expand our universe

Contents

Preface

In the twenty years since I wrote the first edition of *God & the Big Bang*, there have been several significant discoveries in cosmology. Although none of these has compelled me to change any major themes or formulations in the book, I feel that it is worthwhile to revise some of what I have written in light of these new findings.

One of the most significant scientific facts about our universe is that ever since the big bang it has been expanding. As mentioned in chapter 2, this was first demonstrated in 1929 by the astronomer Edwin Hubble. Until nearly the end of the twentieth century, cosmologists assumed that the expansion of the universe was gradually slowing down, due to the combined gravity of all the matter that exists.

However, in 1997 and 1998, two independent scientific teams reported an astounding discovery. Based on their measurements of supernovas (spectacular exploding stars), they demonstrated that for the past several billion years the expansion of the universe has actually been accelerating rather than declerating. In recognition of their work, in 2011 both teams were awarded the Nobel Prize in Physics.

What is causing this unexpected acceleration of the expansion of the universe? According to the most widely accepted hypothesis, it is a mysterious form of energy—a repulsive anti-gravitational force that cosmologists have labeled "dark energy." Several billion years ago, as the density of matter and radiation dropped sufficiently, this dark energy began to dominate the density of the universe. The term "dark energy" resonates with the kabbalistic term for the initial impulse of creation: *botsina de-qardinuta*, "a spark of darkness."

The phrase "dark energy" is not really a physical description, but rather an admission of ignorance, a semantic surrender. This energy has never been directly observed—only inferred from astronomical observations. Over the past fifteen years, two significant satellites were launched: first the WMAP spacecraft (Wilkinson Microwave Anisotropy Probe, operated by NASA), and then the Planck spacecraft (operated by the European Space Agency). These have enabled scientists to determine that dark energy constitutes approximately 68 percent of the total mass or energy of the universe. Only 5 percent of this total is made up of ordinary, atomic matter—constituting galaxies, stars, planets, and everything we can see and touch. Most of the rest (about 27 percent) is composed of another hypothetical substance: an exotic type of matter that does not interact with light and is therefore invisible. Like dark energy, this "dark matter" has never been directly observed, but only inferred—partly from its gravitational effects on ordinary, visible matter. Yet even though we cannot observe this dark matter, billions of dark-matter particles are passing through each of us every second.

In other words, according to contemporary science, 95 percent of what exists in the universe is invisible and its nature unknown. We understand almost nothing about dark matter—and even less about dark energy. It seems that the more we know about the cosmos, the more mysterious it becomes.

Does the big bang, which serves as the scientific creation myth of our culture, have anything to do with God? Can it enrich our lives?

Physicists and theologians often contend that religion and science are two separate realms, each valid within its domain and operating under its own set of rules. The purpose of science is to explore nature, while the purpose of religion is to foster spirituality and ethics. But the question "How did the world come to be?" is vital to both disciplines because it is so fundamental—a question that humans have pondered ever since consciousness evolved to a point where it could reflect on itself and the cosmos.

This book opens with an account of creation according to the theory of the big bang. An overwhelming majority of cosmologists regard the big bang as the most reasonable explanation of the evolution of the universe, "the best approximation to truth that we currently possess."

But the name of this theory, "the big bang," does not convey the awesome nature of the origin of the universe. Furthermore, it is misleading because it implies that matter and energy exploded like a giant firecracker or an immense nuclear bomb in the middle of empty space. Yet according to the theory, space itself is part of the expansion of the universe, and matter is just carried along by the expanding space. So the bomb analogy breaks down. There were no sound waves to make the "bang" audible; the expansion was smooth and continues to this day.

The term "big bang" was coined by a bitter opponent of the theory: the English astronomer and physicist Fred Hoyle. In 1949, Hoyle gave a radio talk for the BBC on twentieth-century cosmology. Detesting the notion that the universe had a beginning, he held a different theory, according to which the universe is eternal. In his talk, Hoyle criticized "the hypothesis that all the matter in the universe was created in one big bang at a particular time in the remote past.... This big bang hypothesis ... is an irrational process that cannot be described in scientific terms." Although it is often claimed that Hoyle intended the phrase pejoratively, he denied this, insisting that he was simply seeking a striking image for the theory he opposed. In any case, the name gradually stuck. The origin of the cosmos has such grandeur, such an effect on our imaginations, that it has bestowed a measure of grandeur on the term "big bang" retroactively.

Still, some scientists and science writers yearned for a different, more evocative name, so a popular astronomy magazine, *Sky and Telescope*, sponsored a contest to find one. Not one of the 13,000 entries impressed the esteemed panel of judges enough to warrant replacing Hoyle's phrase.

Leaving the name aside, how does this contemporary creation story affect—or challenge—our concept of God? Can it help us discover a spiritual dimension in our lives and recover a sense of wonder? *God & the Big Bang* wrestles with these questions. In conceiving and formulating answers, I have drawn on the insights of traditional Jewish learning, especially the mystical traditions of Kabbalah and Hasidism, as well as contemporary physics and cosmology. I suggest several intriguing parallels, but my purpose is not to prove that thirteenth-century kabbalists knew what cosmologists are now discovering. Rather, in juxtaposing these two distinct approaches—the scientific and the spiritual—I experiment with seeing each in light of the other. I am not trying to synthesize the two, because their unique perspectives should not be collapsed. This book seeks, rather, to bring the two into dialogue.

It is said that science demystifies nature, but scientists on the frontier are awed by the elegance and harmony of nature. As science reveals the secrets of the universe and deciphers the cosmic code, it evokes wonder. What science shows us about the evolution of our universe and our selves is as awe-inspiring as the accounts in Genesis or the Kabbalah. Let me give two examples: the color of the sky and the force of gravity.

Why is the sky blue? Among the wavelengths of light in the sun's spectrum, blue oscillates at the highest frequency and is, therefore, scattered effectively by molecules of air in our atmosphere. Because the blue light is bouncing off air in all directions, the sky turns blue. To me, this is more amazing than ancient Mesopotamian and biblical beliefs that the sky is blue because of all the water up there.

Of the fundamental forces of nature, gravity is the one that most obviously affects our everday experience. Gravity was conceived by Newton as a force exerted by objects *in space*, but Einstein showed that it is a property *of space*: the curvature, or "warping," of spacetime. Imagine stepping on a trampoline. Your mass causes a depression in the stretchy fabric—and similarly with space. If you roll a ball past the warp at your feet, it curves toward your mass. The heavier you are, the more you bend space. As summed up by the physicist John Archibald Wheeler, "Spacetime tells matter how to move; matter tells spacetime how to curve."

On a cosmic scale, the force of gravity is now overpowered by dark energy, which is accelerating the expansion of the universe. However, our own galaxy is not expanding—it is simply being carried along by the overall expansion—so within our galaxy and solar system, and on our planet, gravity still wields its power unimpeded.

Gravity is relatively weak between objects that have small masses, but it grows stronger as the objects grow in mass. So something as large as the Earth exerts a mighty gravitational force. When you throw a ball up into the air, gravity quickly brings it back down to earth—unless you're Superman, able to throw it at a speed of 25,000 mph, fast enough that it can escape our planet's tenacious pull.

As for the Earth, it is being drawn by gravity toward the massive sun. Fortunately, our planet's forward motion (and inertia) counterbalance the sun's gravitational attraction, keeping us in orbit around our star and preventing us from falling into its fires.

The sun, meanwhile—along with the whole solar system—orbits through our Milky Way galaxy and around the massive black hole at the center of the galaxy. And the entire Milky Way orbits with other galaxies in what is known as the Local Group, which itself feels the gravitational pull of the Virgo supercluster of galaxies.

According to the law of gravity, every particle in the universe attracts every other particle. So every object in the universe feels the gravitational pull of all other objects (although the farther away an object is, the less its gravitational pull). Gravity is constantly trying to tug everything together, drawing each separate thing toward the original unity of the cosmic seed before it sprouted in the big bang. Gravity conveys a yearning for oneness.

God is a name that we give to the oneness of it all. The theme of God as oneness is a recurring motif in this book. In part 2, "God, Self, and Cosmos," I explore the tension between this expansive view of God and the traditional, personal God. I also discuss the link between the concept of a personal God and the notion of self. According to Kabbalah, the world exists and we have individual consciousness only because the infinite God has withdrawn Itself from a single point of Its infinity, thereby making room for finite being. We exist individually because we have lost oneness through a process the kabbalists call "the breaking of the vessels." Similarly, contemporary physics speaks of "broken symmetry," through which the initial unified state of being shatters, eventually generating the diversity of galaxies, stars, planets, and life. Physicists search for the symmetry hidden within the tangle of everyday reality. They dream of finding equations that link the apparently distinct forces of nature. Spiritual search, too, in its own way, charts a course through multiplicity toward oneness.

But is oneness livable? In part 3, "Torah and Wisdom," I outline a spiritual path that derives from Jewish tradition while remaining open to the wisdom of other faiths and the insights of science. I also touch on the problem of evil and describe the Jewish mystical technique of transforming "the evil impulse."

The book ends with a brief discussion of the fate of the universe and with reflections on our more immediate future.

Where do we fit in the cosmic scheme of things? Earth, our precious little ball of rock 4.5 billion years old, circles the sun once a year. Our entire solar system revolves around the black hole at the core of the Milky Way galaxy once every 250 million years. Our sun is an inconspicuous star, one of at least 100 billion in our spiral galaxy. Our galaxy is one of at least 100 billion galaxies in the observable universe. Andromeda, our closest neighbor, lies two million light years away. The two of us—Andromeda and the Milky Way—are members of the Local Group, on the outskirts of the Virgo supercluster. Beyond lie so many clusters and superclusters that it takes volumes just to catalog them. They all appear to be arranged into gigantic domains that resemble the cells of a sponge.

We are infinitesimal, yet part of something vast. Becoming aware of this, we strive to comprehend the entirety. On this quest, spirituality and science are two tools of understanding. Their approaches to the question of our origins are distinct and should not be confused; each is valid in its domain. Occasionally, though, their insights resonate with each other. By sensing these resonances, our understanding deepens, nourished by mind and heart.

Acknowledgments

I want to thank a number of people for their help and advice. I benefited greatly from the insights and feedback of the cosmologist Joel Primack and the astrophysicist Howard Smith. Among the others who helped me refine my ideas are David Biale, Arnold Eisen, Elaine Markson, Andrew Porter, Robert J. Russell, and Howard Simon. I was stimulated by my students at the Graduate Theological Union and in a course I taught at Stanford University.

In the course of my writing, I heard that biblical scholar Richard Elliott Friedman was also working on the theme of the big bang and Kabbalah as part of his book *The Disappearance of God*. We met, talked, exchanged manuscripts, and became friends. I appreciate his deep learning, wit, and warmth.

Arthur Magida, editorial director of Jewish Lights, deserves a special thanks from me and the reader for his keen eye and his generous efforts to clarify my writing.

I am grateful to Stuart M. Matlins, publisher of Jewish Lights, for his enthusiasm and encouragement and for his vision.

Finally, an offering of thanks to Hana, for her love and support, for her soothing and inspiring presence.

Part One

THE BIG BANG

1. In the Beginning

In the beginning was the big bang, fourteen billion years ago. The primordial vacuum was devoid of matter, but not really empty. Rather, it was in a state of minimum energy, pregnant with potential, teeming with virtual particles. Through a quantum fluctuation, a sort of bubble, in this vacuum, there emerged a hot, dense seed, much smaller than a proton, yet containing all the mass and energy of our universe. In far less than a trillionth of a second, this seed cooled and expanded wildly, faster than the speed of light, inflating into the size of a grapefruit.

During this inflation, the potential mass and energy could not yet manifest as particles; space was expanding too fast for any particles to congeal out of the vacuum. But as the expansion slowed down, energy latent in the vacuum precipitated as particles and antiparticles. These annihilated each other, except for one in a billion particles, which survived to become the building blocks of matter. The annihilation gave birth to a flood of energy, generating the radiation of the big bang. The ball of the universe continued expanding—and has never stopped.

In its first few seconds, our universe was an undifferentiated soup of matter and radiation. It took a few minutes for things to cool down enough for protons and neutrons to form into the simple nuclei of heavy hydrogen and helium. But it was still far too hot for entire, stable atoms to hold together. Within about ten minutes, the production of nuclei stopped.

For the next several hundred thousand years, the universe was somewhat like the interior of a star, filled with subatomic particles and photons (radiant particles of energy). In this early eon, the photons occupied a range of the electromagnetic spectrum beyond what the human eye can see. The radiation was so turbulent and energetic that electrons could not stick to nuclei to form full-fledged atoms. As soon as an atom began to form, it was immediately ripped apart by radiation. Frenzied photons collided with free electrons, traveling only an infinitesimal distance before the next collision absorbed or scattered them. Since no photon could escape, the mixture of radiation and particles was essentially opaque, like a thick, impenetrable fog.

As the universe continued expanding, its temperature and energy gradually fell. After 400,000 years of cooling, when the temperature reached 3,000° Kelvin (4,940° F), a transition occurred. Having lost a critical amount of energy, the photons could no longer tear away electrons from circling around nuclei. Relieved of the photons' constant harassment, the electrons were now free for the first time to settle into orbit around the nuclei, forming stable atoms of hydrogen and helium, which, millions of years later, would grow into stars. Meanwhile, with unattached electrons no longer available, the photons were also freed, able to travel great distances without colliding with an electron and being scattered or absorbed. The photons broke away from the building blocks of matter and became visible, flying through space in all directions. Matter and radiation had decoupled, and the universe turned transparent. This is the moment of "Let there be light!," *Yehi or.*

THE ECHO OF THE BIG BANG

Ever since, for fourteen billion years, radiation has pervaded space. As the universe expanded and cooled, the radiation's frequency continued decreasing, below the visible and infrared ranges, and into the microwave region. This cosmic background radiation is the residual heat signature of the big bang—its faint, persistent echo.

In the late 1940s, the physicists Ralph Alpher and Robert Herman theorized that this echo was still circulating through the universe. But it was not until 1964 that it was discovered—by accident, serendipitously—in New Jersey.

Two radio astronomers, Arno Penzias and Robert Wilson, were scanning the heavens with a new twenty-foot horn-shaped reflector antenna at Bell Telephone Laboratories in Holmdel, New Jersey. Wherever they turned the antenna, they picked up a mysterious background hiss. They first attributed this to electrical static caused by contaminants inside the horn antenna—maybe accumulated pigeon droppings. They dismantled and cleaned the throat of the antenna, but the static persisted. Meanwhile, in Princeton, about twenty-five miles west of Holmdel, the physicist Robert Dicke and his colleagues were designing an experiment to detect the cosmic residue of the big bang. Penzias heard about the Princeton work and contacted Dicke. Upon hearing of Penzias's and Wilson's discovery, Dicke turned to his colleagues and said, "Well, boys, we've been scooped!" The puzzling hiss was indeed the cosmic background radiation, the echo of the big bang. Its temperature was close to what Alpher and Herman had predicted years earlier, about three degrees (Kelvin) above absolute zero.

The cosmic background radiation offers us an image of the universe 400,000 years after the big bang, the last time radiation interacted with matter. In 1989, the Cosmic Background Explorer satellite (COBE) was launched to analyze the structure of the background radiation in microscopic detail. Three years later, the COBE team announced that they had detected slight variations in the temperature of the radiation—the imprints of ancient ripples in the fabric of spacetime. These variations, caused by slight differences in the density of the universe in different regions, reveal that matter was not uniformly distributed: here and there hydrogen and helium were already being tugged together by invisible "dark matter," beginning to assume a structure that would evolve into galaxies and stars. The large-scale structures in the universe today—the clusters of thousands of galaxies—illuminate these ancient ripples "like glitter tossed on invisible lines of glue." For eons, in areas of our universe that were slightly denser than average, gravitation had its effect, slowing down the expansion.

Imagine a galaxy beginning to form: First, dark matter and ordinary matter (hydrogen and helium) form a halo. Within this halo, the hydrogen and helium cool further, falling toward the center. Gradually growing in density, these glimmering clouds of gas form into stars. Within each star, more helium is formed by the fusion of hydrogen, generating energy and fueling the star. After the core of the star has been mostly converted into helium, the star becomes a red giant and begins fusing helium into still heavier nuclei such as carbon, silicon, and oxygen, generating further energy.

If the star is of medium size (more massive than our sun), after eventually turning into a red giant it will blow off its outer layers to form a gas cloud called a planetary nebula. These nebulae scatter elements such as carbon, nitrogen, and oxygen into space.

If the star is originally massive enough (at least eight times as massive as our sun), it will eventually convert most of the material in its core into iron. Because iron has such a tightly bound nucleus, it becomes impossible to extract any energy from it by further nucleosynthesis. At this point, the massive star is doomed. The energy generated by fusion had supported the star's outer layers, countering the force of gravity; without this support, the star collapses. As the entire mass of the star plunges toward its core, a burst of energy rebounds from the center as a huge shock wave. The star explodes, in the violent process forging even heavier elements, such as copper and silver. These, along with the lighter elements from the exploding core, are spewed into space. (Other heavy elements, such as gold, platinum, and uranium, may be ejected from the merger of very compact stars known as neutron stars.) Over eons, these various elements recycle themselves into new solar systems. Our solar system is one example of this recycling, a mix of matter produced by cycles of stars forming and exploding. We, along with our planet and everything on it, are literally made of stardust.

STIRRING THE PRIMORDIAL SOUP

Ten billion years A.B.T. (after the beginning of time), in a spiral galaxy called the Milky Way, about two-thirds of the way out from the galaxy's center, an immense cloud of gas and dust started to contract. (The gas consisted of hydrogen and helium; the dust, of interstellar particles such as carbon and silicon atoms.) As the cloud collapsed toward its center, its rate of spin increased. This is similar to the spinning speed of a figure skater increasing as she pulls her arms inward. The centrifugal force of the spin flattened the cloud into a disk. Counter to this outward thrust, the attraction of gravity pulled most of the matter to the central region, eventually forming the sun. Within the flattening disk, particles not drawn into the center were able to cluster and cool into chunks of material, which in turn collided to form planets. Attracted by the gravity of the sun, each planet eventually settled into an orbit around the solar center, although the early solar system likely had episodes of major rearrangments.

The planet we call Earth took shape and began cooling down about 4.5 billion years ago. By about a billion years later, various microorganisms had developed. Exactly how, no one knows. But we do know that Earth's early atmosphere was composed of hydrogen, water vapor, carbon dioxide, and simple gases such as ammonia and methane. In such a climate, organic compounds may have synthesized spontaneously. Scientists have replicated the Earth's infant atmosphere by blending molecules of its components and exposing the mixture to ultraviolet light (which was stronger when the Earth was young) and electric sparks, approximating the action of lightning. The result: Amino acids assembled themselves, along with other organic molecules. Or perhaps life drifted to Earth in the form of spores from another solar system in our galaxy.

However life began, all its forms share similar genetic codes and can be traced back to a common ancestor; all living beings are cousins. We humans like to think of ourselves as the pinnacle of creation, and it is true that we are the most complicated things that we know of in the universe. We are composed of about ten trillion cells. The retina of the human eye alone contains three million cells, each connected to the brain by an individual wire in the cable of the optic nerve. Our brain contains 100 billion cells, linked by over 100 trillion synaptic connections. Yet, we are part of the evolutionary process, descended from single-cell organisms that lived 3.5 billion years ago. We have evolved through an intricate, unrepeatable combination of chance mutation and natural selection. Our species—*Homo sapiens*—developed in Africa, splitting away from the chimpanzee line about seven million years ago. We still share with the chimps 98 percent of our active genes. If you'll pardon the expression, we are an improved ape.

BEFORE THE BIG BANG

The big bang is a theory, not a fact. To cosmologists, it offers the most convincing explanation of the evolution of the universe, "the best approximation to truth that we currently possess." The scientific consensus is that the big bang theory is correct within its specific domain: the evolution of our universe from perhaps one-billionth of a second after its origins up to the present. Whatever happened before that first fraction of a second lies beyond the limits of the current theory. The term "big bang" suggests a definite beginning a finite time ago, but the theory does not extend that far. In that sense, the first line of this chapter is somewhat misleading. It would have been more accurate to say: "One-billionth of a second after 'the beginning' was the big bang," since the ultimate origin of the universe is still unfathomed.

Beginning in 1979, cosmologist Alan Guth hypothesized and then developed the idea of cosmic inflation. According to this revolutionary theory, for an infinitesimal fraction of a second after "the beginning," space expanded exponentially, perhaps by a factor of 10^{50}. Following this, the universe continued to expand, but much less rapidly. The newborn universe would have had only tiny variations in density caused by quantum fluctuations, but inflation would have enlarged these into the significant variations that existed after 400,000 years. These variations went on to seed the galaxies.

Over the past several decades, Guth's idea has come to dominate theoretical models of the early universe. One particular version of the inflation theory was formulated by Stanford cosmologist Andrei Linde, portraying a universe that reproduces itself continually, attaining immortality. To Linde, our universe is one of countless baby universes, or what he calls "bubbles." The initial conditions in each of these bubbles differ, and diverse kinds of elementary particles interact in unimagined ways. Perhaps different laws of physics apply in each. They may even contain a different number of dimensions of spacetime.

Not all the domains inflate into large bubbles. Those that do, like ours, dominate the volume of the universe and sprout other bubbles in a perpetual chain reaction. The entire universe is a tree of life, a cluster of bubbles attached to each other, growing exponentially in time. Each baby universe is born in what can be considered a big bang—or should we say a little bang?—a fluctuation of the vacuum followed by inflation. Each world begins somewhere in the past and ends somewhere in the future, but the evolution of the entire universe has no end. As to the beginning of the entire universe, Linde's theory allows for an initial bang, but does not require it. Flirting with religious language, Linde suggests that God "created a universe that has been unceasingly producing different universes of all possible types." In the course of its evolution, the universe realizes all possibilities: "Our cosmic home grows, fluctuates and eternally reproduces itself in all possible forms, as if adjusting itself for all possible types of life it can support. The performance is still going on, and it will continue eternally. In different parts of the universe, different observers see its endless variations."

If this is correct, perhaps we should translate the opening words of Genesis not as "In *the* beginning ...," but "In *a* beginning, God created ..." In fact, this could represent a more literal rendering of the original Hebrew: *Be-Reshit*, "In a beginning."

But what happened *before* this beginning?

The first big bang theorist was an obscure Belgian priest and mathematician named Georges Lemaître. In 1931, he proposed that the eruption of "a primordial atom" had given birth to the universe. His theory assumed that the universe emerged out of an infinitely small point of space packed with infinitely dense matter—what physicists call a "singularity." At a singularity, gravity, too, is infinite. The image is mind-boggling, but its depiction of a primordial instant harmonizes with traditional religious belief regarding a definite beginning of the universe. In fact, the Catholic Church endorsed the big bang model in 1951, claiming it accorded with the Bible.

Scientists, meanwhile, sought to demonstrate accordance between the expansion of the universe from a singularity and Einstein's theory of relativity. In the 1970s, physicists Stephen Hawking and Roger Penrose succeeded in doing just that. Later, however, Hawking theorized that in the initial stages of the universe the singularity disappears.

In Hawking's universe, time and space together constitute a four-dimensional foam of spacetime, finite in size yet without boundary or edge. If this seems hard to visualize, don't worry: It's not hard, it's impossible. But start, Hawking suggests, by picturing the two-dimensional surface of the Earth. This surface is finite, but has no edge or boundary: Sail as far as you can, and you won't fall off. Now add a third dimension of space and then a fourth dimension: time. The resulting spacetime has no boundary, no singularity. So it is meaningless to speak about what happened at the boundary, and the notion of a beginning becomes irrelevant.

Let's return to Earth. Stand anywhere on the equator and head north. Eventually, after adding several layers of clothes, you find yourself at the North Pole. You can't go any further north; "north" loses its meaning. Similarly, in the very early universe, the dimension of time becomes harder and harder to define. What we call "time" had a beginning, but that does not mean spacetime has an edge, just as the surface of the Earth has no edge at the North Pole.

We can imagine time stretching back forever, even before the universe existed. But time is simply something that enables us to label events *in* the universe. It is a parameter. Where such a parameter begins is artificial; it doesn't correspond to the edge of reality. Time is defined only *within* the universe. Outside of spacetime, before the beginning of the universe, time has no meaning. Asking what happened before the universe began is like searching for a point one degree north of the North Pole: It's simply not defined. Instead of conceiving of the universe as being created or coming to an end, we should realize that it just *is*.

According to Hawking, time itself began at the moment of the big bang. In confining time within the universe, Hawking follows Philo of Alexandria (the first-century Hellenistic Jewish philosopher) and Saint Augustine (the fifth-century Church father). The former suggested that time began after creation, with the start of motion; the latter concluded that God created time: "What did God do before He made heaven and earth? I do not answer as one did merrily: 'He was preparing hell for those who ask such questions.' For at no time had God not made anything because time itself was made by God."

Philo and Augustine believed in a God who creates the universe and inaugurates time. But for Hawking, the universe is completely self-contained, without boundary or edge and with no external first cause. "What place, then," he asks, "for a creator?"

Science has no consensus on the ultimate origin. Some theories espouse a well-defined beginning; others, like Hawking's, do not. But both suggest a radically new reading of Genesis. If God spoke the world into being, the divine language is energy; the alphabet, elementary particles; God's grammar, the laws of nature. Many scientists have sensed a spiritual dimension in the search for these laws. For Einstein, discerning the laws of nature was a way to discover how God thinks.

But does the universe have a purpose? Is there meaning to our existence? Why should we live ethically? Here, cosmology cannot help us very much. Darwin intensifies our problem. Are we different from other animals? Can we transcend violence and savagery? As the wife of an Anglican bishop remarked, upon hearing of Darwin's theory: "Descended from apes! My dear, let us hope that it is not true; but if it is, let us pray that it will not become generally known." Her comment echoes the fear that knowing the true nature of our ancestors threatens to unravel the social fabric.

MYTH AND MEANING IN OUR LIVES

We have lost our myth. A myth is a story, imagined or true, that helps us make our experience comprehensible by offering a construction of reality. It is a powerful narrative that wrests order from chaos. We are not content to see events as unconnected, as inexplicable. We crave to understand some underlying order in the world. A myth tells us why things are the way they are and where they came from. Such an account is not only comfortable, assuring, and socially useful; it is essential. Without a myth, there is no meaning or purpose to life. Myths do more than explain. They guide mental processes, conditioning how we think, even how we perceive. Myths come to life by serving as models for human behavior. On Friday evening, as our family begins Shabbat, I sometimes imagine God, having created the world in one very packed week, finally taking a break. "God rested and was refreshed," *Shavat va-yinnafash*. This mythical image enables me to pause, to slow down and appreciate creation. By observing Shabbat, I am imitating the divine. Order reemerges out of the impending chaos of life.

But what do we do when the myths of tradition have been undone, when the God of the Bible seems so unbelievable? Is there really someone "up there" in control, charting the course of history, reaching down to rescue those in need, tallying up our good and bad deeds for reward and punishment? Many people have shed the security of traditional belief; they are more likely to experience a gaping, aching void than the satisfying fullness of God's presence. If they believe in anything, perhaps it's science and technology. And what does science provide in exchange for this belief? Progress in every field except for one: the ultimate meaning of life. Some scientists insist that there is no meaning. As one leading physicist has written, "The more we know about the universe, the more it is evident that it is pointless and meaningless."

Science is not the only challenge to believing in God. Human suffering also corrodes faith, leaving us suspended and alone. As Job's suffering led him to question whether God is just, the Holocaust has led many to wonder whether God is dead. But to say "God is dead" is really to make a statement about human beings: It means that the traditional way of conceiving of God no longer works. For after Auschwitz, how can we speak of a caring, compassionate, personal God?

Through the lens of belief, Christians, Muslims, and Jews have gazed at their own history, searching for traces of a divine plan. The Jewish people, in particular, have interpreted their historical traumas as divine punishment. By helping them cope with an imperiled existence, this strategy enabled them to salvage sense from suffering. Biblical prophets linked the destruction of Solomon's Temple to the people's moral failings. Yohanan ben Zakkai, one of the founders of rabbinic Judaism, similarly explained the fall of the Second Temple. The liturgy for the three Jewish pilgrimage festivals includes the lament: "On account of our sins, we were exiled from our country and banished far from our land; so we cannot go make pilgrimage to worship You." Some still invoke this principle to explain the Holocaust. The sixth Lubavitcher *rebbe*, Joseph Isaac Schneerson, for instance, explained that Hitler was God's instrument for chastising the Jews, who had abandoned the ways of Torah. This notion is at once traditional and obscene. Its vulgar Christian analogue—officially repudiated, but still encountered—is: "The Christ-killers got what was coming to them." Most Jews and Christians recoil from such interpretations of the Holocaust. But there are only two alternatives: Either God does not exist—or She is very different than we have thought.

Part Two

GOD, SELF, AND COSMOS

2. Oneness and Nothingness

In its own way, the big bang is a contemporary creation story. Energy turns into matter, which turns back into energy. There is no precise plan for creation worked out in advance. By an intricate and unrepeatable combination of chance and necessity, humanity has evolved over billions of years from and alongside countless other forms of life. Ultimately, our evolutionary history is uplifting: It enables us to see that we are part of a wholeness, a oneness.

To be "religious" means, in the words of a contemporary physicist, to have an intuitive feeling of the unity of the cosmos. Mystics and poets have perceived this oneness, as did the philosopher Heraclitus, who wrote: "When you have listened not to me but to the *Logos*, it is wise to agree that all things are one." This oneness is grounded in scientific fact: We are made of the same stuff as all of creation. Everything that is, was, or will be started off together as one infinitesimal point: the cosmic seed.

Life has since branched out, but this should not blind us to its underlying unity. Every living thing is composed of molecules; molecules are composed of atoms; atoms are composed of electrons (which seem not to be composed of anything smaller) and protons and neutrons, which are composed of three quarks each. The quarks are held together by gluons, so called because they enable the quarks to "glue on" to each other. Yet, if we simply reduce everything to the least common denominator—the simplest elementary particles yet discovered—we miss the delicious differences between things: blue jays and white poodles, eucalyptus and acacia, Koreans and Hasidim. When I am bicycling with my children, Michaella and Gavriel, what brings me a smile is not their sameness—the fact that they are composed of electrons and quarks like everything else—but their inimitable shrieks of joy as they race down the hill, their hair blowing untamed in the wind. No subatomic analysis can explain the wonder of personality, the unique pattern each of us creates, dancing and stumbling through life.

The deepest marvel is the unity *in* diversity, the vast array of material manifestations of energy. Becoming aware of the multifaceted unity can help us learn how to live in harmony with other human beings and with all beings, with all our fellow transformations of energy and matter.

THE POWER OF NAMING

If the big bang is our new creation myth, the story that explains how the universe began, then who is God?

"God" is a name we give to the oneness of it all.

The act of naming is quintessentially human. In Genesis, Adam names all the animals. Developing this thought, the rabbis imagine God passing all the animals in front of Adam, asking him, "What is this one called? And this one?" Adam responds, "Ox, camel, donkey, horse." The first human being then provides his own name. Finally, he provides God with a name: *YHVH.*

God is the oneness of the cosmos. But the name "God" is a label we attach to this oneness—the ultimate, all-inclusive name. In fact, this is exactly how Jews traditionally refer to God: *Ha-Shem,* "The Name."

The very act of naming—on the one hand so bold and powerful—betrays, on the other hand, our linguistic and mental limitations. By naming things, we control them. Or we try to. Naming enables us to call other people and call upon them: command, cajole, implore them. By defining things, we bestow or impose order on the welter that surrounds us. But as we define and classify things, we confine them—and we confine our understanding. The very meaningfulness of our names constricts the reality we are naming.

The names that Adam, Eve, and their countless descendants have assigned to things are useful and necessary, but misleading. We cannot function in this world without names and labels; yet we cannot perceive all that is there if we remain entranced by names, if we do not venture beyond them. My wife's name, Hana, is deeply meaningful to me because it reminds me of what I feel for her, which is beyond words. If I want to see her, I can speak her name: "Hana, could you come here for a minute?" But the moment she comes into the room, her name dissolves in her presence.

As I compose this paragraph, I am looking out the window at a tree. My eye follows a branch and focuses on a leaf. "Leaf."

The name is mentally satisfying. I have found the appropriate label; I know *what* I am seeing. But the appropriateness of the name lulls me into thinking that there really is a separate object called a "leaf," as if the leaf were not part of a continuum: blade-veins-stem-stipule-twig-branch-limb-bough-trunk-root.

So the name "leaf" is misleading. Maybe I should just stick with "tree"? But is there really a separate, self-contained thing I can call by that name? Down below, the roots absorb water and minerals from the soil. Up above, the chlorophyll in the leaves traps and stores the energy of sunlight. The leaf is not separate from the tree; the tree is not separate from the earth and atmosphere. If I pause and reflect, I realize that nothing is entirely separate from anything else. A faint memory of the cosmic seed lingers.

We need names to navigate through life, but those very names obscure the flowing continuum. Behind each handy name is a teeming reality that resists our neat definitions. If this is true of the names we assign to the hundreds of thousands of things of this world, how much more so with our names for God, the oneness of it all.

THE GOD BEYOND GOD

How can you name oneness? How can you name the Unnameable? The Jewish mystical tradition, Kabbalah, offers a number of possibilities. One is *Ein Sof,* which translates literally as, "there is no end." *Ein Sof* is the Infinite, the God beyond God. This name sounds so different from the personal divine names that populate the texts of the tradition: *YHVH, Elohim, Shaddai,* the Holy One, blessed be He. As an anonymous kabbalist observed, neither the Bible nor the Talmud even hints at *Ein Sof.* This remark is both obvious and revealing, an acknowledgment of the radical originality of this mystical formula.

By calling God *Ein Sof*, Jewish mystics imply that *everything* is divine. The kabbalist Moses Cordovero, writing in the sixteenth century, put it this way: "The essence of divinity is found in every single thing—nothing but It exists. Since It causes every thing to be, no thing can live by anything else. It enlivens them. *Ein Sof* exists in each existent. Do not say, 'This is a stone and not God.' Perish the thought! Rather, all existence is God, and the stone is a thing pervaded by divinity."

There is nothing but *Ein Sof*. Even a stone in a field, even a slab of concrete in a downtown parking lot, is an expression of divine energy. The entire world is God in myriad forms and disguises.

The name *Ein Sof* opens with a negative: *Ein*, "there is no." This accords with the view of the philosopher Moses Maimonides that it is more accurate to say what God is *not* than what God *is*. Calling God "powerful" conjures up the image of a muscleman. Calling God "wise" makes us think of a sage. Better to say that God is neither "weak" nor "stupid." Even the bland statement "God exists" is misleading because divine existence is unlike anything that humans can conceive. A more precise formulation is: God "exists but not through existence."

The best theology, in Maimonides' view, is negative theology. "Know that the description of God by means of negations is the correct description, a description that is not affected by an indulgence in facile language. Negative attributes conduct the mind toward the utmost reach that one may attain in the apprehension of God. You come nearer to the apprehension of God with every increase in negations."

The kabbalists adopt Maimonides' negative style of theology—and take it to an extreme. Among their names for God, *Ein Sof* is the most famous, but not the most radical. Having carved away all that is false, they discover a paradoxical name: *Ayin*, "Nothingness." We encounter this bizarre term among Christian mystics as well: John Scotus Erigena, writing in Latin, calls God *nihil;* Meister Eckhart, in German, *nihts*; St. John of the Cross, in Spanish, *nada*.

To call God "Nothingness" does not mean that God does not exist. Rather, it conveys the idea that God is no thing: God animates all things and cannot be contained by any of them. In the words of a fourteenth-century kabbalist, David ben Abraham ha-Lavan, "Nothingness (*Ayin*) is more existent than all the being of the world. But since it is simple, and all simple things are complex compared with its simplicity, in comparison it is called *Ayin*." David ben Abraham's Christian contemporary, Meister Eckhart, concurred: "God's nothingness fills the entire world; His something though is nowhere."

Ayin is a name for the nameless. The paradox is that *Ayin* embraces "nothing" *and* "everything." This nothingness is oneness: undifferentiated, overwhelming the distinctions between things. God is the oneness that is no particular thing. No thingness. Nothingness with a capital *N*.

This mystical nothingness is neither empty nor barren; it is fertile and overflowing, engendering the myriad forms of life. Medieval philosophers—Jewish, Christian, and Muslim—had taught that God created the world "out of nothing." The mystics turn this formula on its head, reinterpreting it to mean that the universe emanated from divine nothingness. Similarly, as we have seen, cosmologists speak of the quantum vacuum, teeming with potential, engendering the cosmic seed. This vacuum is anything but empty—a seething froth of virtual particles, constantly appearing and disappearing. According to quantum field theory, pairs of these virtual particles, one positive and one negative, appear together in the primordial vacuum, move apart, then come together again and annihilate each other. Even if cooled to absolute zero, the vacuum still shimmers with a residual hidden energy: what physicists call "zero-point energy."

As we will discover, Kabbalah also describes a primordial vacuum, at the heart of *Ein Sof*. Like its quantum counterpart, this vacuum is not absolutely empty, but rather coated with a trace of divine light.

How did the universe emerge out of prolific nothingness? According to Kabbalah and classical big bang theory, this transition was marked by a single point. Recall the singularity: an infinitely dense point in spacetime. Like *Ayin*, a singularity is both destructive and creative. Anything falling into a singularity merges with it and loses its identity, while energy emerging from a singularity can become anything. The everyday laws of physics do not apply to that split second in which energy or mass emerges.

According to the thirteenth-century kabbalist Moses de León, "The beginning of existence is the secret concealed point. This is the beginning of all the hidden things, which spread out from there and emanate, according to their species. From a single point you can extend the dimensions of all things. Similarly, when the concealed arouses itself to exist, at first it brings into being something the size of the point of a needle; from there it generates everything."

In Kabbalah, this point is identified with *Hokhmah*, divine "wisdom," which emerges from nothingness into being. The *Zohar*, the masterpiece of Kabbalah, opens by disclosing the origin of the point:

> A spark of darkness flashed
> within the concealed of the concealed,
> from the head of Infinity,
> a cluster of vapor forming in formlessness....
> Under the impact of splitting,
> a single, concealed, supernal point shone.
> Beyond that point, nothing is known,
> so it is called Beginning.

As emanation proceeds, as God begins to unfold, the point expands into a circle. Similarly, ever since the big bang, our universe has been expanding in all directions. We know it is still expanding thanks to Edwin Hubble, who measured the speed at which other galaxies are moving away from us. In 1929, Hubble determined that the farther a galaxy is from us, the faster it is moving away.

THE "RAISIN THEORY" OF THE UNIVERSE

Imagine you are a galaxy. If this sounds too grandiose, then imagine you are a raisin in a batch of dough. As the dough is baked, it expands into a raisin cake. As it expands, you look out at the other raisins. The more distant ones have more expanding dough between them and thus move apart faster, while nearby raisins have less dough separating them and move more slowly. From our vantage point here in the Milky Way galaxy, the other galaxies are moving away from us, their velocity increasing with distance. It's not that the universe is expanding *within* space. Space itself is expanding, and the galaxies, like the raisins, remain the same size. As mentioned in the preface, for the past several billion years the expansion of the universe has been accelerating.

The most dramatic consequence of Hubble's discovery is what it tells us about the origin of our universe. Just play the Hubble tape in reverse: If the universe is now expanding, that means it was once much smaller. How small? According to classical big bang theory, if we go back far enough in time and retrace the paths of the galaxies and their formation, the entire mass-energy of the universe contracts into the size of a singularity—the infinitesimal point from which the cosmos flashed into existence.

In Kabbalah, the expansion is pictured in a number of ways, for example, as the outcome of a powerful divine breath: "When a glassblower wants to produce glassware, he takes an iron blowpipe, hollow as a reed from one end to the other, and dips it into molten glass in a crucible. Then he places the tip of the pipe in his mouth and blows, and his breath passes through the pipe to the molten glass attached to the other end. From the power of his blowing, the glass expands and turns into a vessel—large or small, long or wide, spherical or rectangular, whatever the artisan desires. So God, great, mighty, and awesome, powerfully breathed out a breath, and cosmic space expanded to the boundary determined by divine wisdom, until God said, 'Enough!'"

This heavenly command is based on an imaginative midrash on the divine name *Shaddai*: the one who said, "*Dai*!" ("Enough!"). Another kabbalist, Shim'on Lavi, understands expansion as part of the secret rhythm of creation:

> With the appearance of the light, the universe expanded.
> With the concealment of the light, the things that exist were created in all their variety.

> This is the mystery of the act of Creation.
> One who understands will understand.

When light flashed forth, time and space began. But the early universe was an undifferentiated soup of energy and matter. How did matter emerge from the stew? The mystic writes that the light was concealed. A scientist would say that energy congealed. Matter is frozen energy. No nucleus or atom could form until some of the energy cooled down enough that it could be bound and bundled into stable particles of matter.

Einstein discovered the equivalence of mass and energy. Ultimately, matter is not distinct from energy, but simply energy that has temporarily assumed a particular pattern. Matter is energy in a tangible form; both are different states of a single continuum, different names for two forms of the same thing. Compare this to Cordovero's remark: "Do not say, 'This is a stone and not God.'" Divine energy pervades all material existence.

Quantum physics demonstrates how teasingly difficult it is to pinpoint the building blocks of matter. According to Heisenberg's uncertainty principle, which lies at the heart of quantum theory, the position and velocity of subatomic particles cannot both be determined simultaneously. If you try to measure precisely where an electron is, its speed becomes indeterminable and vice versa. The more precisely you measure either position or velocity, the more indeterminable the other one becomes. Because of the inherent indeterminability, quantum physicists cannot establish both definite results for such an observation. Rather, they consider various possible outcomes and assign each one a probability. The very nature of a particle is uncertain; it behaves as both particle and wave. In a sense, it is neither until it is observed. The observer participates in the construction of reality.

Like the physicist, the mystic, too, is fascinated by the intimate relation of matter and energy, though the mystical description is composed in a different key. Material existence emerges out of *Ayin,* "no-thingness," the pool of divine energy. Ultimately, the world is not other than God, for this energy is concealed within all forms of being. Were it not concealed, there could be no individual existence; everything would dissolve back into oneness or nothingness. The light, paradoxically, reveals itself only by being concealed. As Cordovero puts it, "When powerful light is concealed and clothed in a garment, it is revealed. Though concealed, the light is actually revealed, for were it not concealed, it could not be revealed. This is like wishing to gaze at the dazzling sun. Its dazzle conceals it, for you cannot look at its overwhelming brilliance. Yet when you conceal it—looking at it through screens—you can see and not be harmed. So it is with emanation: By concealing and clothing itself, it reveals itself."

We cannot behold the infinite, but its power is displayed through everything that exists. Creation is a form of revelation. The underlying oneness is not apparent, but it is real.

COUNTERING IDOLATRY

God is the oneness of matter and energy, the process through which one is transformed into the other, the nothingness that embraces both. Nothingness may seem like a shocking name for God, but it follows logically from Judaism's command against idolatry, the second of the Ten Commandments: "You shall have no other gods beside Me." It's rare, these days, to find people actually bowing down to graven images, but we constantly constrict God within our various mental images, thinking that He or She has a particular form. God is worshiped as Mother or

Father, as Provider, Judge, Ruler. These metaphors can be very effective: reassuring us, inspiring us to act ethically or spiritually. But when the metaphor hardens into a fixed image, we lose more than we have gained. Our awareness of God becomes limited to the particular image we focus on; we reduce—and even desecrate—the infinite nature of God. As a Spanish rabbi said in the twelfth century, whoever thinks that God has an image is fashioning idols and bowing down to them. Idolatry is as much a mental as a physical act.

Ayin is an antidote to idolatry. It explodes the notion that God is an object. *Ayin* forces us to surrender our comfortable, confining images: It melts them down. This "*Nichts* of the Jews," writes the poet Henry Vaughan, exposes "the naked divinity without a cover." But how can we think or speak of God without images and conceptions? We can't. Even *Ayin* is a concept. The images it evokes may be vast: a limitless ocean, the expanse of outer space. But they are images nonetheless. The value of *Ayin* is that it helps us understand the relative nature of all images of God, including *Ayin* itself.

Images and names of God enable us to approach the divine, but they can't quite get us there. They keep us at a safe distance. These pictures and words indicate the reality, but cannot convey it. To encounter God directly, we need to leave names and images aside. We must renounce the idolatry of worshiping any particular image, of relying too much on any particular name. On the threshold of experience, we are challenged to let go of words, to attune ourselves to the "sound of sheer silence," *qol demamah daqah*.

3. The Personal God—and Beyond

The God of the Jewish Bible is treated almost as a person. Only rarely described as feminine, God often appears as a heavenly patriarch, compassionate yet irascible. Called by such names as *Shaddai, El, Elohim, YHVH,* this God is so personal that He is in love with the people of Israel and jealous of any other gods trying to lure away His beloved. Conceiving of God as a person implies a relationship, but also implies a gap, since the divine personality is assumed to be separate from us and from all nature. God did not emanate and become the world; God majestically *spoke* the world into being. As Saint Augustine formulated the view, "The works of creation were made from nothing by You, not of You." Intimacy with God forms the core of spiritual life: loving God with heart, soul, and force. But our relationship depends on the separateness of divine and human identities. Though I am created in the image of God, we relate to each other *as other.*

Without a personal God, there is no possibility of relationship with the divine. How can I relate to the boundless? The boundless includes and subsumes me, along with everything else. Nothing is separate from infinity.

The God of the Jewish mystics is both personal and impersonal. In the personal mode, God becomes even more anthropomorphic than in the Bible. Yet the roots of divine personality are embedded in nothingness and infinity. The personal God is born out of *Ayin.*

THE TEN *SEFIROT*: THE COSMIC TREE OF LIFE

The mystics are reticent about *Ein Sof,* which is fitting when speaking of the infinite, but they indulge in describing the ten *sefirot:* the various stages of God's inner life and the dynamics of divine personality. The *sefirot* depict God more graphically than we find anywhere in the Bible or the Talmud. Not only does God feel, respond, and act through the *sefirot,* but they constitute an androgynous divine body, complete with arms, legs, and sexual organs. Here, God is both He and She, and the union of the divine couple conveys blessing to the world.

We have already encountered the first *sefirah*: *Ayin.* The nothingness of *Ayin* is undifferentiated oneness, roughly the same as infinity. In fact, some kabbalists treat *Ein Sof* and *Ayin* as one and the same. From here, the other nine *sefirot* emerge.

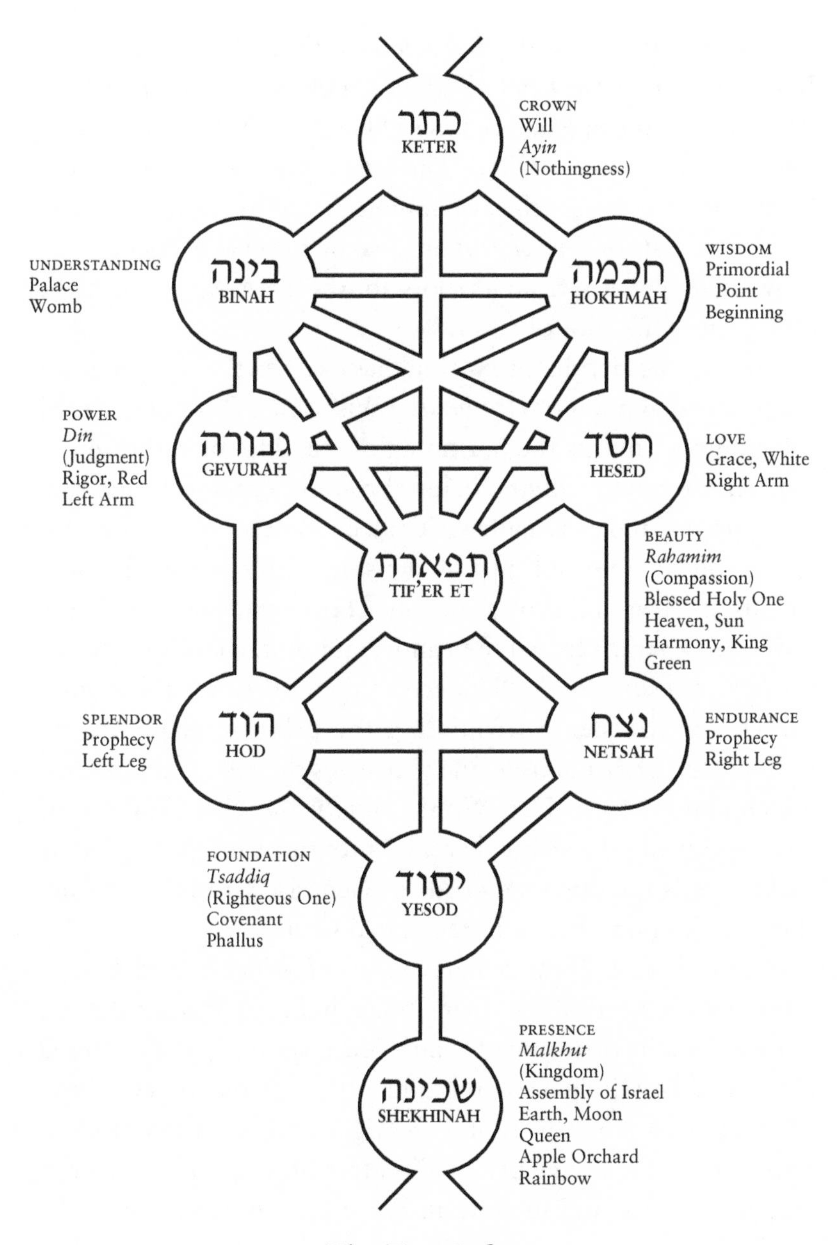

The Ten *Sefirot*

More commonly, the first *sefirah* is called *Keter,* "Crown." It is the crown on the head of *Adam Qadmon,* primordial Adam. According to the opening chapter of Genesis, the human being is created in the image of God. The *sefirot* are the divine original of that image, the mythical paragon of the human being, our archetypal nature. The *sefirot* are also pictured as a cosmic tree, growing downward from its roots above—from *Keter,* the highest *sefirah,* "the root of roots."

From the depths of Nothingness shines the next *sefirah*: the primordial point of *Hokhmah* "Wisdom." This point, called "Beginning," marks the beginning of creation, or rather, emanation: the flow of divine being. From here, the sefirotic tree branches out and eventually yields all existence. From this point, space and time unfold, just as the singularity of the big bang signals the beginning of spacetime. From the singularity, the universe expands; from the primordial point, a circle expands: the next *sefirah, Binah,* "Understanding." *Binah* is the womb, the Divine Mother. Surrounding the primordial point, the cosmic seed of *Hokhmah,* She conceives the rest of the *sefirot,* which emerge from Her. Within *Binah,* the "personality" of God begins to take shape; here, infinity turns into God. As the *Zohar* reads the opening words of Genesis, "With Beginning, through Wisdom, the Infinite created God."

The three highest *sefirot* (*Keter, Hokhmah,* and *Binah*) represent the head of the divine body and are considered more hidden than the offspring of *Binah*. She gives birth first to *Hesed* (Love) and *Din* (Judgment), the latter often referred to as *Gevurah* (Power). This pair constitutes the right and left arms of God, two poles of the divine personality: free-flowing love and strict judgment, grace and limitation. For proper functioning of the world, both are essential; ideally, a balance is achieved, which is symbolized by the central *sefirah, Tif'eret* (Beauty), also called *Rahamim* (Compassion). If judgment is not softened by love, then it lashes out and threatens to destroy life. Here lies the origin of evil, which is called *Sitra Ahra,* "the other side." The demonic is rooted in the divine.

Tif'eret is the masculine trunk of the sefirotic body. He is called "Heaven," "Sun," "King," and "the Holy One, blessed be He," the standard rabbinic name for God. The next two *sefirot* are *Netsah* (Endurance) and *Hod* (Splendor). They form the right and left legs of the divine body. Relatively little is said about them, except that they are the source of prophecy. *Yesod* (Foundation) is the ninth *sefirah* and represents the phallus, the procreative life force of the universe. *Yesod* is the *axis mundi,* the cosmic pillar. The light and power of the preceding *sefirot* are channelled through *Yesod* to the tenth *sefirah, Malkhut* (Kingdom), or *Shekhinah* (Presence).

THE RETURN OF THE GODDESS

Shekhinah is the name for God's immanence in the Talmud and Midrash. But in the Kabbalah, *Shekhinah* becomes a full-fledged She: the feminine side of God, daughter of *Binah*, bride of *Tif'eret. Shekhinah* is "the secret of the possible," receiving emanation from above and engendering varieties of life below. She complements Her masculine partner, the Holy One, blessed be He, mollifying His occasional outbursts. The joining of *Shekhinah* and *Tif'eret*—the feminine and masculine halves of God—becomes the focus of spiritual life. Human beings stimulate the divine union by acting ethically and religiously, thereby assuring an abundant flow of blessing to the world. Human marriage symbolizes and actualizes divine marriage, while the evening of Sabbath turns into a weekly celebration of the cosmic wedding—and the ideal time for human lovers to unite.

Shekhinah represents a partial, yet significant, corrective to patriarchal religion. God's maleness was no Jewish invention. The transition from Goddess to God, from dominant female deities to dominant male deities, occurred long before the composition of the earliest books of the Bible. To the ancient Hindus, Indra, the warrior god, reigned supreme; to the Greeks, it was Zeus; to the Germanic tribes, Thor. The Bible gradually elevates a tribal warrior god, *YHVH,* to the status of the transcendent monotheistic deity of the universe. The prophets rail against *YHVH*'s rivals among the Canaanite fertility gods and goddesses, which proves that the Israelites engaged in such forbidden worship.

As demonstrated by archaeological evidence, the cult of Asherah (the Canaanite Mother Goddess) flourished in ancient Israel. It's not just that there were Israelites who worshiped the Canaanite Baal and Asherah instead of the one God of Israel. Rather, some Israelites worshiped *YHVH* and Asherah together as a divine couple. For example, at an archaeological site about twenty-five miles south of Jerusalem, an inscription has been found dating from the eighth century B.C.E. that speaks of the blessings granted by "*YHVH* and His Asherah."

This syncretism was rejected by the biblical authors and prophets, and the Goddess was eliminated from the official religion of Israel. Yet somehow, in Kabbalah, She reemerges as *Shekhinah*. The Goddess has become kosher!

Apparently, the ancient feminine nature of God could not be supressed forever. The new flowering of *Shekhinah* testifies to the Goddess's enduring hold on human consciousness. Maimonides, the great medieval philosopher, had sought to cleanse God of any lingering anthropomorphisms, to bury myth once and for all. But in the century following his death, *Shekhinah* emerged in all Her feminine glory. The renowned scholar of Jewish mysticism Gershom Scholem has called this "the revenge of myth." Once Kabbalah resurrected the feminine aspect of God, over the following centuries She became immensely popular among the masses. Clearly, She answered a deep religious and psychological need: Her intimacy and accessibility balanced the transcendent, remote patriarchal God. The feminine nature of God is thus both new and ancient. *Shekhinah* represents the God who dwells within. God is not just beyond us, but all around us and within us.

The kabbalists were not proto-feminists. *Shekhinah,* relegated to the last rung of the sefirotic ladder, is subservient to the Holy One, blessed be He. *Shekhinah* is generally passive and receptive; She transmits the flow of divine emanation, but is said to have "nothing at all of Her own." Contemporary feminists are suspicious of this domesticated Goddess: Some prefer a God beyond gender; other are more attracted to *Shekhinah*'s independent and demonic shadow, Lilith, who *was* a proto-feminist. As Adam's first wife, she refused to lie beneath him while making love and fled, causing mischief ever since. Clearly, the kabbalistic image of *Shekhinah* is a masculine product, fashioned by men for men. Yet She is a leading character in the masterpiece of Kabbalah, the *Zohar,* which devotes more space to Her than to any other *sefirah*. She fascinated the *Zohar*'s male authors, who realized something radically obvious: God cannot be limited to a masculine description. Today, seven hundred years later, the *Zohar*'s theological critique is still widely unheeded.

TOOLS FOR MEDITATION

The multiple personality and mythical imagery of the sefirotic God is stunning. The kabbalists insist that their figures of speech should not be taken literally and that these are organic symbols of a spiritual reality beyond normal comprehension. Sefirotic descriptions are intended to convey something of the beyond, and becoming fixated on the image itself defeats the purpose.

The theory of *Ein Sof* and the *sefirot* seems dualistic: an impersonal, infinite oneness versus a personal, embodied God who interacts not only with the world and humanity, but also between Himself and Herself. Moreover, the multiplicity of the *sefirot* seems to compromise, if not to threaten, the absolute oneness of God. As the iconoclastic thirteenth-century kabbalist Abraham Abulafia remarked, some who believe in the *sefirot* have outdone Catholic adherents of the trinity and turned God into ten! Kabbalists maintain that *Ein Sof* and the *sefirot*—the impersonal and the personal—form a unity, "like a flame joined to a burning coal." "They are Its name, and It is they." "It is they, and they are It." Though the *sefirot* appear independent and multiple, they are essentially one: waves on the surface of *Ein Sof*.

The personality of God emerges from and leads back to *Ein Sof*, the boundless. By contemplating the *sefirot,* one explores the emotional and psychological texture of God's various personal qualities, such as love, fear, and compassion. The *sefirot* are not mere abstract concepts or divine metaphors deployed by kabbalistic authors in some intricate literary game. They are tools for meditation. Each serves as a focus of visualization, disclosing depths of archetypal personality within God and the seeker. Yet, contemplation is not an end in itself. In the words of

Moses Cordovero, "the essence of the divine image is action." The mystic is called upon to imitate God, to embody the *sefirot,* to be gracious, rigorous, and compassionate. Here, for example, is how Cordovero suggested embodying *Hesed,* grace:

> Desire the well-being of your fellow creature, eyeing his good fortune benevolently. Let his honor be as precious to you as your own, for you and your fellow are one and the same. This is why we are commanded: "Love your neighbor as yourself." Desire what is right for your fellow; never denigrate him or wish for his disgrace, suffering, or ruin. Feel as bad for such suffering as if it were you own. Similarly, rejoice over another's good fortune as if you were basking in it.

"It is impossible to conduct yourself according to these qualities constantly," Cordovero conceded. "Accustom yourself to them little by little. The essential quality to attain, the key to them all, is humility," which is the human counterpart of the highest *sefirah,* Nothingness.

The *sefirot* function as an ethical training course, a systematic way of translating spiritual ideals into concrete action in the world. Yet at the same time, the *sefirot* constitute a mystical path, a ladder of ascent back to the One. Glimpsing its origin, the human soul yearns to retrace its route, from here to infinity. This yearning fuels a spiritual quest, a journey of endless discovery. Now, the *sefirot* serve as a map of consciousness. Climbing and probing, the mystic uncovers dimensions of being.

The mystical and the ethical reinforce one another, through a rhythm of going deep within and living mindfully in the material world. You can experience spiritual and psychological wholeness by imitating and integrating the sefirotic attributes. "When you cleave to the *sefirot,* the divine holy spirit enters into you, into every sensation and every movement." The path is not easy. Unexplored shadows of our psyche must be encountered, not evaded. By knowing and withstanding the dark underside of wisdom, the spiritual seeker is refined.

Near the top of the sefirotic ladder, approaching the source of emanation, meditation reaches *Binah,* the Divine Mother. She is also called *Teshuvah,* "Return." In Her, the ego returns to the womb of being. *Binah* cannot be held in thought, as indicated by another one of Her names: Who. This primal question serves as a focus of contemplation: "Who am I?" The questioning eventually yields nothing that can be grasped, just an intuitive flash illuminating and disappearing, as sunbeams play on the surface of water. As Moses de León wrote,

> Thought reveals itself only through contemplating a little without content, contemplating sheer spirit. The contemplation is imperfect: You understand—then you lose what you have understood. Like pondering a thought: The light of that thought suddenly darkens, vanishes; then it returns and shines—and vanishes again. No one can understand the content of that light. It is like the light that appears when water ripples in a bowl: shining here, suddenly disappearing, then reappearing somewhere else. You think that you have grasped the light, when suddenly it escapes, radiating elsewhere. You pursue it, hoping to catch it—but you cannot. Yet you cannot bring yourself to leave. You keep pursuing it.
>
> It is the same with the beginning of emanation. As you begin to contemplate it, it vanishes, then reappears; you

understand—and it disappears. Even though you do not grasp it, do not despair. The source is still emanating, spreading.

BEYOND THE PERSONAL GOD

In the depths of *Binah* lies *Hokhmah* (Wisdom). The seeker is nourished from this primal sphere, which cannot be known consciously; it can only be absorbed. In the words of Isaac the Blind, one of the founders of Kabbalah, who lived at the turn of the thirteenth century, "The inner, subtle essences can be contemplated only by sucking, not by knowing." Beyond *Hokhmah* is the nothingness of *Keter,* the annihilation of thought. In this ultimate *sefirah,* human consciousness expands and expands, finally dissolving into infinity.

Ein Sof is the God beyond God, beyond the personal God. Whereas the *sefirot* embody the male and female personality of God, *Ein Sof* is transpersonal. While the *sefirot* fulfill the human need for a personal relationship with God, in the background beckons the infinite. The *sefirot* may constitute a set of divine archetypes, in whose ideal image we are said to be created, but is God really like that? Ultimately, according to the *Zohar*, one discovers that the *sefirot* have no existence independent of *Ein Sof*: "Upon contemplating…, there is nothing but the Supernal Lamp." The personal God appears to exist only "from our perspective," *mi-sitra di-lan.*

How can anyone relate to something as abstract as infinity or nothingness? Ask a typical believing Jew, Christian, or Muslim: "What is your concept of God?" The answer will probably describe a very personal God, a God who can almost be pictured. What is this deep need for a personal God? Why do so many people believe in this type of divine being? Part of the answer is that we know we will someday die. Painfully aware of our own mortality, we yearn for the comfort of a cosmic parent who will always be there for us. Even if the sun will burn out five billion years from now, there must be *something*, we think, that endures forever, some ultimate One on whom we can rely. By believing in a personal God who always was, is, and will be, we convince ourselves that somehow we, too, can achieve a form of immortality. We hunger for eternity.

It's comforting to feel that someone out there is watching over us and caring for us, that we have a Mother or Father in Heaven. But this belief is constantly contradicted by the suffering that stains our lives. If we believe in a personal God who knows and cares, then how do we explain earthquakes and devastating fires and the tragic death of children? Such messy facts threaten to ruin our pristine portrait of God.

We see and experience many things that mock the notion of a personal God. Yet, many of us still maintain a personal relationship with Her or Him. This relationship is based not only on our awareness of mortality or our need to be cared about, but also on a more fundamental connection between the self and a personal God. As we will now see, these two notions are intertwined.

4. Self and God

We have not always had a self, so why do we need one now? A six-month-old baby makes no clear distinction between itself and the world, between subject and object. With no defined self, the baby's "self" seems to pass into objects, and objects into it. These objects are impermanent: perceptual pictures that appear, dissolve, and reappear. Whether or not we remember, there was a time early in each of our lives when we didn't know how to differentiate between what is inside and outside our skin, between what is "self" and "other." Separateness was not yet a category for us; the baby has a self only to the extent that it is "other" than mommy. Then one day, we spoke the magic word "I" and everything changed.

The Talmud relates that Adam originally extended "from one end of the universe to the other." The first human being had no clear-cut sense of a separate self. His consciousness was unbounded: He was one with the cosmos. Only after eating from the Tree of Knowledge of Good and Evil did Adam begin to make distinctions and evaluations. As he become aware of a limited self, his cosmic consciousness contracted and he was reduced to ordinary, mortal dimensions.

We could not survive or function in the world without a separate sense of self, without an ego. If we didn't recognize the boundary between our body and the outside world, we would soon be killed by water, fire, a wild animal, or an automobile. Humans, of course, are not unique in this regard. No living thing can survive unless it distinguishes between self and other. In a sense, consciousness first appeared on Earth 3.5 billion years ago, with the advent of microorganisms. They, too, make a distinction between inside and outside, between me and you, us and them. They exhibit a peculiar, very rudimentary consciousness of self.

Picture yourself a few billion years ago, as a microbe in the primeval oceans. You survive and grow by eating organic molecules dissolved in the rich ocean water. You are eating molecules that make up other beings. You and they are composed of the same molecules, so you have to be careful not to eat yourself! You must make some fine distinctions. If you can't discriminate between "me" and "you," if you can't control your digestive enzymes, then you'll leave fewer offspring—or none. As a microbe, you do not think or feel, but you behave *as if* you had wants, needs, preferences, drives, instincts.

Every cell in the human body can distinguish between itself and others. Those that cannot will succumb to disease and death. But when we speak of human consciousness, we mean more than this. We like to think that our self-awareness distinguishes us from all other creatures, rendering us unique. If a fish, a cat, a dog, or a bird catches sight of itself in a mirror, it does not know that it is seeing itself. It perceives the image as simply another member of its species. Humans are different. At about the age of two, a child understands that the image in the mirror is its own. But are we so unique? Psychologists have discovered that adult chimpanzees can develop mirror self-awareness in a few days. Having figured it out, they use the mirror to preen, and to examine inaccessible parts of themselves, looking over their shoulders to view their backs, for example. Orangutans and dolphins also exhibit mirror self-awareness.

THE DAWN OF CONSCIOUSNESS

What makes us different? It is not memory; many animals have this. It is not the ability to reason; chimpanzees and macaques reason at a fairly elementary level. Worms and even one-celled protozoa can be taught to run a simple maze. All mammals know how to play. Various animals, especially the primates, exhibit friendship, altruism, love, fidelity, courage, intelligence, invention, curiosity, and forethought. Humans differ from other species, but only by degree, and mostly because enhancing certain talents for invention, forethought, language, and intelligence has enabled us to understand and change the world in extraordinary ways.

But unavoidably, we are animals—and like other animals, we evolved with the talent to sense what was needed for survival and to focus on those aspects of the environment essential for staying alive while ignoring everything else. The human brain assimilates just one-trillionth of the information that reaches the eye. Our senses, brain, and nervous system initially developed because they helped our prehuman ancestors discriminate between safe and dangerous stimuli, to dodge falling objects, to recognize faces, to avoid predators. Consciousness dawned for a very practical reason: A conscious—and self-conscious—creature was more adept at hunting and gathering, mating and child-rearing since a conscious brain assesses probability and risk.

As evolution proceeded, brains grew larger relative to the weight of the body. Birds have relatively bigger brains than fish; mammals than birds; primates than other mammals. Humans weigh only a little more than a chimpanzee, but our brain is four times more massive.

The size of the brain is only part of the story. The human brain reached its current size about 150,000 years ago, before the development of language, cooking, or agriculture. Our tremendous advance, especially over the last ten thousand years, is due mostly to harnessing the plasticity of the brain in radically new ways—by creating something like software to enhance its underlying powers. After a long history of natural selection, humans redesigned their brains through self-manipulation. By exercising our mental powers, we have learned not only how to survive, but how to thrive. We created art, science, philosophy, values, civilization. Science has given us a fairly clear picture of how the universe and life evolved. We have identified hidden patterns of energy all around us and within us; we have cracked the genetic code and the cosmic code. We have discovered the cosmic background radiation: the echo of the big bang. We know enough to realize how little we know, especially about ourselves.

What exactly is consciousness? Our experience of consciousness may be subjective, but it is a biological feature of the brain, caused by the interactions between billions of individual cells. Every tenth of a second, a stream of these cells forms, then falls away. The cooperation among these cells gives rise to consciousness. While it appears to be continuous, in fact, consciousness has distinct gaps. It allows for no permanent, abiding, essential self. Although we assume that there is a separate, conscious entity at our core, this is a fiction, one that is wondrous and necessary. It is not that we are less than a self: We are more, since we are part of the oneness of the cosmos. The left hemisphere of our brain persuaded us that we were otherwise. This marvelous piece of architecture, which excels at analysis and thought, established a boundary between its own contrived ego—its personal identity—and the rest of the universe.

Over our lifetime, in collaboration with family and friends, we have woven a story about ourselves that defines who we are. The ego cannot be understood or expressed except in relation to an audience, and this audience's responses—real or imagined—continually shape our telling of the story. We do not consciously and deliberately figure out what narratives to tell and how to tell them. For the most part, we don't spin our tales: They spin us.

These streams of narrative issue forth *as if* from a single source. To those around us, it seems that a unified being has authored the story, that there is a center of narrative gravity. This apparent center, this apparent self, is an enormously helpful simplification, but it is an abstraction, not a "thing" in the brain. Though fictional, it is remarkably robust, almost tangible. Our rich narrative is organized around a hub of self-representation. The hub isn't a self: It's a *representation* of a self. It gathers and organizes the information on the topic of "me," just as other structures in the brain keep track of information on food or on the way back home.

FASHIONING GOD IN OUR IMAGE

What does all this have to do with a personal God? According to Genesis, Adam and Eve were evicted from the Garden of Eden because they ate from the Tree of Knowledge of Good and Evil. As they learned to make distinctions and judgments, innocence vanished, yielding alienation—which ever since has proven to be an unavoidable consequence of growing up and living in the world. The *Zohar,* commenting on this fateful passage, asks, "Who divorced whom? Did the blessed Holy One divorce Adam, or not?"

Or *not?* What is the alternative? The *Zohar* doesn't spell out the secret immediately, either because the thought is too radical or so that it can surface slowly and astoundingly in the mind of the reader. The secret is that God did *not* divorce Adam. Rather, Adam divorced God. More precisely, Adam divorced *Shekhinah,* the feminine divine presence. This suggests several possible meanings. Adam ruined the marriage between the feminine and masculine halves of God. Instead of contemplating the union of *Shekhinah* with Her divine husband (the Holy One, blessed be He), Adam split them apart.

Or, Adam divorced himself from the feminine, a mistake his male descendants have repeated ever since and only recently begun to realize.

Or—and this is what interests me here—Adam severed himself from the oneness of it all and became identified with ego. Exclaiming "I am," he divorced himself from God.

Before the emergence of the self, there is no "other." The mental construct of "self" sets the stage for all relationship—with other human beings, with other objects, with a personal God. God as oneness is there from the start. But as the ego splits itself off from oneness, it invents and discovers "others," all of which are shattered fragments of oneness. The ego affirms its separateness by conceiving the other. Its most spectacular conception, its greatest self-affirmation, is a personal God.

The ego and its personal God are dependent on each other. They are also mutually reinforcing. The impermanent self needs the anchor of a personal God since belief in such a God fulfills a deep need of the ego: that there is someone "out there" who believes in me and believes that I exist as a separate self. Belief in a personal God, then, confirms my separateness. Conversely, this personal God needs our affirmation. This mutuality was not lost on a second-century sage, Rabbi Shim'on bar Yohai, who interpreted a verse in Isaiah, "'You are My witnesses,' says *YHVH*, 'and I am God,'" as meaning: "When you are My witnesses, I am God. When you are not My witnesses, I am not, as it were, God."

Ego and a personal God participate in an ancient, secret covenant—sharing a pact of mutual preservation, conspiring together against a oneness that would overwhelm their separateness. They reinforce each other's otherness. Without a personal God, it is easy to lose any sense of being an integrated self, of being something more than just a mental tangle of drives, desires, and disappointments. By declaring in the *Shema* that God is one—"Hear, O Israel: *YHVH* our God, *YHVH* is One!"—I insist, too, on my own singularity.

But God's oneness should not be limited to its personal aspect. It is also transpersonal, enfolding and overwhelming my separate sense of self. While chanting the opening words of the *Shema*, I affirm that I am part of God's oneness. I acknowledge that the separateness of my ego and the personality of God are illusory mirror images.

Who is made in whose image? Through eons of evolution, out of the oneness of it all, we have been fashioned, eventually emerging as a conscious self. Through us, the oneness has become conscious of itself. And, in a knee-jerk reaction, a personal God has been fashioned by us, projected in our own image. To this God, we attribute our creation.

The ego is a marvelous fiction. We do not possess it: It possesses us. Not having an ego is unimaginable, an unbearable thought. There are good reasons for preserving the myth of the self as a particular, concrete thing rather than an abstraction. Without this myth, society would be doomed, which is why we spend so much time and energy constructing a self that can accept moral responsibility. We need the ego. Madness and anarchy are not attractive alternatives. To put it less dramatically, when the phone rings, we answer it and identify "who we are." Without such an "identity," chaos rushes in.

In the overall scheme of life, a personal God is appropriate for the ego. As long as I see my self as a separate entity, I can relate only to a God who stabilizes my fragile ego and lends it meaning. But the otherness of God and self is maintained only by a tension that links them together. Though the ego insists on keeping its distance, it is drawn across the divide. In the words of Psalms, "As a deer pants for brooks of water, so my being longs for You, O God." Through a clouded lens, the self dimly sees that it is part of something greater. It gropes for the oneness that remains just out of reach—a oneness addressed as *You,* but often obscured by projection, by being fashioned in our image.

The part yearns to rejoin the whole. Letting go of mental images, the self discovers it is no longer merely an isolated fragment, but rather a unique expression of the Self of the universe. As the self becomes aware of the Self, personal consciousness tastes the oneness. In the words of another psalm, "Taste and see that God is good." Yet God—the Self behind all selves—is not a passive object of our budding spiritual awareness. By evolving through spacetime, by organizing Itself into the wondrous variety of existence, God grows and learns endlessly, discovering awareness through each of us—God's countless, inimitable selves.

5. Cosmic Hide-and-Seek

As God was planning to create Adam and Eve, He said, "Let us make a human in our image" (Genesis 1:26). Why does God speak in the first person plural? Why not "Let Me make a human in My image"? To the rabbis of the Midrash, "let us" evokes a heavenly court in which a retinue of angels surrounds God. God proposes that they join Him in creating human beings. Some angels oppose the idea because mortals are prone to evil and failure. Elsewhere in the Midrash, "let us" is interpreted as referring to other conceivable partners or consultants of God, such as the preexistent souls of the righteous or the preexistent, archetypal Torah. The *Zohar* offers a different view: "Us" refers to the entire cosmos, "showing that God is everything."

Fashioned in the image of oneness, we reflect oneness, but we also refract it through our prism of individuality. Each of us is a fraction of infinity. But since infinity cannot be quantified, no matter how it is divided and subdivided, even a fraction of infinity is infinite. So even in our fractured state, we manifest infinity. Through being who we are, each of us expresses the cosmic oneness distinctively. I am a unique creation, yet my most basic physical substance—my quarks and electrons—are identical with those of an antelope, a redwood, a distant star. I am made of stardust.

But, unlike the rest of creation, I *know* that I am made of stardust, that I am one with the cosmos and simultaneously a separate, conscious person. Yet, we usually take the simple route and convince ourselves that we are nothing but a unique and separate ego. Only occasionally, when we feel extremely safe and secure, do we possibly let down our defenses for a moment, a timeless moment. We then loosen the knot of self, and invite someone else into our domain. We surrender to love.

Love is what we feel when we become aware of our oneness with what we thought was separate from us: a person, a place, a thing, an idea. This oneness and love is God. A bit of *gimatriyya* (numerology) provides a mnemonic device for this equation. The Hebrew word for "one" is *ehad,* and the word for "love" is *ahavah*. According to the technique of *gimatriyya,* each Hebrew letter has a numerical value. The *gimatriyya* of *ehad* is the sum of its individual letters: 1 *(alef)* + 8 *(het)* + 4 *(dalet)* = 13. The *gimatriyya* of *ahavah* is 1 *(alef)* + 5 *(he)* + 2 *(bet)* + 5 *(he)* = 13. Oneness and love are equivalent. Together, *ehad* and *ahavah* add up to 26. This is the same numerical value assigned to the holiest divine name, *YHVH,* the sum of whose letters (10+5+6+5) also equals 26. God is oneness and love.

To be religious is to cultivate an appreciation of oneness, to be open to the possibility of love. Of course, you can't be in love with everyone and everything or feel oneness all the time. You'd never get anything done. Worse than that, you wouldn't survive for long. Driving down the road, you don't want to become one with anything moving in the opposite lane. The challenge is to balance oneness and separateness, to acknowledge both realms.

ORIGINATING IN THE BIG BANG

We cannot survive without the self. But if constrained by a narrow sense of self, we cannot be fulfilled. There are moments when the self uncovers its vast ground of being, its interface with all that exists. Mystics have no monopoly on such moments. When we experience an overwhelming feeling of love—for daughter, son, parent, partner, or anyone—the self is unsealed. Love enables the ego to let go of itself and touch another. In nature, you can feel oneness beckon. The wide expanses invite us to slow down and gaze at the clouds, the horizon, the ocean. There are various ways in which we transcend our limited sense of self—by delving into a meaningful project, a profound book, or a worthy cause, or simply by pausing and witnessing a flower or a face. By engaging something beyond our ego, we drop the burden of self-consciousness.

You can't tell when this will happen. Usually we act as if we were autonomous, independent beings. But occasionally—at a waterfall, on a walk, hugging someone we love—we glimpse a trace of infinity. Something inside us remembers the oneness.

As members of the cosmos, we derive from the big bang. Every quark within each of us originated in that primal beginning. As we have seen, the law of gravity implies that everything that exists is drawn to the primordial oneness of the cosmic seed.

The universe is still reverberating from the big bang, as evidenced by cosmic background radiation, which has been coursing through space for nearly fourteen billion years. Our highly developed mind may be able to perceive the origin of the universe. Through science, the deductive mind observes and gropes its way back as far as possible and as close as possible to the beginning of time and space. Through contemplation, the meditative mind observes and gropes *its* way back. The two approaches, while not the same, are complementary paths from our limited, human vantage point to the beginning.

Why does the scientific approach work? Granted that much remains unknown, how is it that we can piece things together to the extent that we can? As Einstein wondered, "The most incomprehensible thing about the universe is that it is comprehensible." Because the evolution of our brains followed the workings of natural law, they resonate with the organization of the natural world. We, too, emerged from the cosmic order, so our cognitive capabilities reflect that order. As children of the cosmos, we not only explore the nature of our universe, but glimpse its laws.

Einstein's discovery of the law of relativity was stimulated by his fascination with light. As a teenager, he conducted the following thought experiment: What if you were running after a light wave and attained its speed? The light wave should stop. But, as Einstein noted, "Something like that does not seem to exist." This mental experiment became the nucleus of his theory of relativity. He eventually concluded that not only can you never catch up to a light wave, you can't even get close. No matter how fast you go, you will always be separated from it by the speed of light. This speed—unique, ultimate, and absolute—is unattainable by any material object. To reach the speed of light, you would have to dematerialize.

IMAGINING YOU ARE LIGHT

Einstein was not alone in imagining light and contemplating its secrets. Describing meditation, one thirteenth-century kabbalist recommended the following:

> Whatever one implants firmly in the mind becomes the essential thing. So if you pray and offer a blessing to God, or if you wish your intention to be true, imagine you are light. All around you—in every corner and on every side—is light. Turn to your right, and you will find shining light; to your left, splendor, a radiant light. Between them, up above, the light of the Presence. Surrounding that, the light of life. Above it all, a crown of light—crowning the aspirations of thought, illumining the paths of imagination, spreading the radiance of vision. This light is unfathomable and endless.

Mystical insight, in its own poetic way, illuminates the emergence of world and self, the resonance between mind and nature. The mystic aims to recover the cosmic consciousness of the first days of Eve and Adam, the limitless self that most of us taste so rarely that we forget we have access to it. Azriel of Gerona, a thirteenth-century kabbalist, urged his readers to reclaim intimacy with the divine mind: "Say to Wisdom, 'You are my sister.' Join thought to divine wisdom, so she and he become one."

The human and divine partners are attracted to each other. In the words of Isaac of Akko, a fourteenth-century kabbalist: "The soul will cleave to the divine mind, and the divine mind will cleave to her. For more than the calf wants to suck, the cow wants to suckle. She and the divine mind become one, like pouring a jug of water into a gushing spring: All becomes one."

Fortunately, this meditative experience is rare. It would be hard to sustain in the supermarket and hard to survive while crossing the street—though contemplating Isaac of Akko's final image could magically transform the very mundane act of rinsing dishes. In our normal mode of consciousness, we effortlessly think ourselves into being separate from everything around us. In the words of Abraham Isaac Kook, the twentieth-century rabbi, poet, and mystic, "Usually the mind conceals the divine extremely by imagining that there is a separate mental power that constructs the mental images. But by training yourself to hear the voice of God in everything, the voice reveals itself to your mind as well. Then right in the mind, you discover revelation."

Revelation cannot be conveyed as long as the "I" monopolizes the mental process. The human mind can link up with the divine, if the self is ready for the encounter. The self is not destroyed or erased by the experience, but rather seen through, via the penetrating lens of nothingness.

This nothingness is not nihilistic. It does not imply that everything is meaningless, dismal, or nonexistent. Its point is simply that everything is interdependent and interrelated. My self is part and product of an infinite web of relations: I am Hana's husband; Michaella and Gavriel's father; Hershel and Gustine's son; Debbie, David, and Jonny's brother; my students' teacher; my neighbors' neighbor; an American citizen; a member of the squabbling human family; a child of the cosmos. Ultimately, my identity is not separate from countless strands of being.

FROM HUMILITY TO AWE

We have seen how Kabbalah speculates on the creative nothingness of God. Hasidism, the eighteenth-century movement that revitalized Jewish spirituality, recasts these kabbalistic speculations in a psychological mold. Experiential aspects of *Ayin* become prominent. The emphasis is no longer on the *sefirot,* the inner workings of divinity, but on how to perceive the unity hidden throughout the world and how to transform the ego. One of the founders of Hasidism, Dov Baer, the Maggid ("preacher") of Mezhirech (in the Ukraine), encouraged his followers to look at the self differently. Playing with the Hebrew letters of *aniy,* the word for "I," the Maggid permutes them into *Ayin,* challenging his listeners to reconfigure the ego. Only by attaining awareness of *Ayin* can one imitate and express the boundless nature of God. As Dov Baer wrote, "Think of yourself as *Ayin* and forget yourself totally. Then you can transcend time, rising to the world of thought, where all is equal: life and death, ocean and dry land. This is not the case when you are attached to the material nature of this world. If you think of yourself as something, God cannot be clothed in you, for God is infinite and no vessel can contain God—unless you think of yourself as *Ayin.*"

By seeing through the apparent solidity of self, we shed the illusion of being totally separate from what surrounds us. Of course, to survive and function in the world, we have to respect the boundaries of self, but in moments of insight, love, or laughter, the boundaries expand and the self becomes less isolated and more a part of the "all." To stubbornly defend the idea of a thoroughly independent self is a sign of false pride. True humility involves consciousness of *Ayin,* as indicated by one of Dov Baer's disciples:

> The essence of serving God and of all the *mitzvot* is to attain the state of humility: to understand that all your physical and mental powers and your essential being depend on

> the divine elements within. You are simply a channel for the divine attributes. You attain this humility through the awe of God's vastness, through realizing that "there is no place empty of It." Then you come to the state of *Ayin*, the state of humility. You have no independent self and are contained in the Creator. This is the meaning of the verse [Exodus 3:6]: "Moses hid his face, for he was in awe." Through his experience of awe, Moses attained the hiding of his face: He perceived no independent self. Everything was part of divinity.

At the burning bush, Moses' sense of separateness was consumed. His mask of personality melted away; his self was effaced. But Moses was exceptional. Ego usually persists and finds subtle ways to maintain itself even in a context of nothingness, as illustrated by the following joke:

> One Yom Kippur in a small, crowded synagogue, the rabbi stood in front of the ark, pouring out his heart like water. Having completed a litany of confessions, he added his own spontaneously: "God, I am nothing, absolutely nothing!"
>
> The cantor, witnessing such fervor, approached the ark, too. "God," he called out, "I have led Your children today in prayer. Together, we have offered words to crown You on high. But in truth, I know that I am nothing, absolutely nothing!"
>
> Silence reigned. And then, a simple Jew near the back of the congregation, inspired by the rabbi and cantor, cried out: "God, I am nothing, nothing!" At which, the cantor leaned over to the rabbi and whispered: "Look who thinks he's nothing!"

Nothingness cannot be claimed. To claim it, or to deny it to someone else, is to lose it. Nothingness is not a possession; it is letting go. It surrenders even its own nothingness.

Ayin does not induce a blank stare. On the contrary, it engenders new mental life, through a rhythm of annihilation and fresh thinking. As Dov Baer taught, "Turn away totally from the prior object of thought, toward a place called 'nothingness.' Then a new topic comes to mind. Transformation comes about only by passing through nothingness." In the words of one of his disciples, "When you attain the level of gazing at *Ayin*, your intellect is annihilated. Afterwards, when you return to the intellect, it is filled with emanation." This creative pool of nothingness is the "preconscious" (*qadmut ha-sekhel*), which precedes, surpasses, and inspires both language and thought. Of this, Dov Baer said, "Thought requires the preconscious, which rouses thought to think. This preconscious cannot be grasped. Thought is contained in letters, which are vessels, while the preconscious is beyond the letters, beyond the capacity of the vessels. This is the meaning of the verse: 'Wisdom emerges out of nothingness.'"

It is possible to trace each thought, each word, each material thing back to its source in *Ayin*. Then, the world no longer appears separate from God. According to the Chabad school of Hasidism, "If we perceive the world as existing [independently], that is merely an illusion." "The foundation of the entire Torah is that *Yesh* [the apparently separate "somethingness" of the world] be annihilated into *Ayin*. The purpose of the creation of the worlds from *Ayin* to *Yesh* was that they be transformed from *Yesh* to *Ayin*." This transformation is realized through contemplative action. "In everything they do—even physical acts such as eating—the righteous raise the holy sparks, from the food or anything else. They thus transform *Yesh* into *Ayin*."

WHERE BOUNDARIES DISAPPEAR

This mystical perspective is neither nihilistic nor anarchic. Matter is not destroyed or negated. It is enlivened and revitalized. By becoming aware that energy animates material existence, you increase the flow from the source to its manifestation. As Dov Baer explained,

> When you gaze at an object, you bring blessing to it. By contemplating that object, you realize that it is really absolutely nothing without divinity permeating it. Through this contemplation, you draw greater vitality to that object from the divine source of life, since you bind that thing to absolute *Ayin*, from which all beings have been hewn. On the other hand, if you look at that object and make it into a separate thing, by that very look, the thing is cut off from its divine root and vitality.

World, mind, and self dissolve momentarily in *Ayin* and then reemerge. *Ayin* is not the goal in itself; it is the moment of transformation from being through non-being to new being. This is much like a seed in the earth, disintegrating as it begins to sprout. The natural process of annihilation engenders fresh life. As Dov Baer points out, "When you sow a single seed, it cannot sprout and produce many seeds until its existence is nullified. Then, raised to its root, it can receive more than a single dimension of existence. There in its root the seed becomes the source of many seeds."

Ayin is the root of all things, and "when you bring anything to its root, you can transform it. Each thing must first arrive at the level of *Ayin*; only then can it become something else."

Nothingness embraces all potentiality. Every birth and rebirth navigates the depths of *Ayin*, as when a chick emerges from an egg: For a moment, "it is neither chick nor egg."

How does this relate to the self? As long as the ego refuses to acknowledge its source, to participate in the divine, it mistakes its part for the "all." Fooling itself, the ego imagines that it possesses itself, but what it claims as its own is simply one manifestation of God. In the words of the Hasidic master Menahem Mendel of Kotsk, "The 'I' is a thief in hiding." When this apparently separate self is *ayin*ized, the result is not death, but the emergence of a new human form, a more supple image of the divine. Only when "one's existence is nullified is one called 'human.'"

Ayin underlies and undermines the manifold appearance of the world. The ten thousand things we encounter daily are not independent or fragmented, as they seem. There is an invisible matrix, a swirl generating and recycling being. If you venture into this depth, you must be prepared to surrender what you know and what you are, what you knew and what you were. The ego cannot abide *Ayin*, where, for an eternal moment, boundaries disappear. *Ayin*'s "no" clears everything away, making room for a new "yes," a new *Yesh*.

BREAKING THE VESSELS

In our cosmic game of hide-and-seek, God hides within each of us, within all of creation and throughout spacetime. The Hebrew word for "universe," *olam*, originally meant "eternity," so the word spans all of time and all of space: spacetime. According to the mystics, *olam* derives from the same root as "hiding," *he'lem*. God is disguised as the world, and the purpose of the game of creation is to uncover the divine, to explore the limits of who we are, to actualize God's self-awareness. Our very consciousness is the universe becoming aware of itself, God becoming aware of Itself. When the divine spark within each human creature discovers that it is not separate from the God beyond, the players—or, rather, the Player in all Its guises—is overjoyed. In the words of an early Hasidic writer, Meshullam Feibush Heller, "When we become aware that only God exists, as before Creation, then God receives from us the true delight He hopes for."

Before creation, there was only *Ein Sof*, God as infinity. But if *Ein Sof* pervaded all space, how could there be room for anything other than God? How could the process of divine emanation begin? Pondering such questions in the Galilean city of Safed in the sixteenth century, the kabbalist Isaac Luria concluded that the first act of creation was not emanation, but withdrawal: "Before the creation of the universe, *Ein Sof* withdrew Itself into Its essence, from Itself to Itself within Itself. Within Its essence, It left an empty space, in which It could emanate and create.... When *Ein Sof* withdrew Its presence all around in every direction, It left a vacuum in the middle, surrounded on all sides by the light of *Ein Sof*, empty precisely in the middle."

This is known as *tsimtsum*, literally "contraction," but here implying the withdrawal by which God made room for something other than God. The primordial vacuum carved out by *tsimtsum* was a pregnant void, the site of creation: no more than an infinitesimal point in relation to *Ein Sof*, yet spacious enough to house the cosmos. But the vacuum was not really empty. It retained a trace, a residue of the light of *Ein Sof*—just as the vacuum preceding the big bang was not completely empty, but rather in a state of minimum energy: pregnant with creative potential and virtual particles.

As *Ein Sof* began to unfold, a ray of light was channeled into the vacuum through vessels. Everything went smoothly at first, but some of the vessels, less translucent, could not withstand the power of the light, and they shattered. Most of the light returned to its infinite source, "to the mother's womb." But the rest, falling as sparks along with shards of the shattered vessels, was eventually trapped in material existence. Our task is to liberate these sparks of light and restore them to divinity. As the Egyptian kabbalist Israel Sarug advised, "You should aim to raise those sparks hidden throughout the world, to elevate them to holiness by the power of your soul." By living ethically and spiritually, we raise the sparks and thereby bring about *tiqqun,* the "repair," or mending, of the cosmos.

If the vessels had not broken, our world of multiplicity would not exist. We exist because we have lost oneness. We have forgotten ourselves into existence. Our spiritual task—the object of the cosmic game—is to recover what has been lost, to remember, to raise the sparks. In the words of Dov Baer,

> The act of *shevirah* [breaking] was essential to the existence of the universe. If every object and aspect were still united with the root and as nothingness, none of the worlds would be. If the material world were continually united with the Creator—without any forgetting—all that exists would be nullified, united with the root, with *Ayin.* Considering themselves to be nothing, they would do nothing! So there had to be *shevirah,* causing a forgetting of the root. Now, human beings can initiate action. Through Torah and prayer, we join the root, *Ayin,* thereby raising the sparks of the material world, delighting God. This delight is even greater than continuous delight—as when a father who has

> not seen his son for a long time is finally reunited with him: The father is more overjoyed than if the son had always been with him. The son, too, having not seen his father for such a long time, yearns all the more passionately to be together again.

First, God withdraws, concealing Itself. This leaves a gap of spacetime, an arena for the universe and the self. As God recedes, creation unfolds. Since oneness cannot be contained by any vessel, the vessels break, spilling out variety. Every thing becomes what it is, and we imagine ourselves as separate entities, in need of *tiqqun*.

DISCOVERING THE SYMMETRY OF THE UNIVERSE

Modern cosmology has a theory that parallels the breaking of the vessels: the theory of broken symmetry. Symmetry, which means "the same measure," implies harmonious proportions. An object is symmetrical if it looks the same from different points of view. The human face, for example, has bilateral symmetry. Because there is little substantial difference between the two sides of your face, it looks nearly the same whether viewed directly or in the mirror—with right and left reversed.

Some things have more extensive symmetries than the human face. A snowflake looks the same when viewed from six different directions; so does a cube. A sphere looks the same from any direction. Symmetries are everywhere: in flowers and palm fronds; in pine cones, seashells, and starfish; in music, poetry, sculpture, architecture, kaleidoscopes, weaving, and square dancing. If something looks, sounds, or feels beautiful to us, it probably displays symmetry. Our minds resonate with the hidden symmetries of nature.

But symmetry can be unstable. Picture yourself at an elegant wedding dinner, sitting with other guests around a circular table. Champagne glasses have been placed precisely between each dinner plate and the next: perfect right-left symmetry. A waiter comes and fills the glasses with champagne—and everyone sits, waiting for someone else to lift a glass. You're a little thirsty; realizing that those pink bubbles won't last forever, you decide to take a sip. But which champagne glass should you pick? Not fully versed in the rules of etiquette, you could as easily choose the glass to your left as the one to your right. Either way, as soon as you reach for one or the other, the symmetry is broken. Unless everyone else does exactly what you've done, someone's going to have to reach across the table to get a glass.

Let's take a more mundane example. Imagine that you're holding a handful of pencils, just snugly enough that they stand straight on their points on a surface. Now let go. For a moment, the pencils remain balanced—rotationally symmetrical. Looking down from above, you see a perfect circle of pencil erasers. But, of course, the symmetry is quickly broken, as the pencils fall into a jumble.

The pencils are a metaphor for the universe. The jumble of fallen pencils is the universe today, while the symmetrical bundle is the universe in its original state. One of the great challenges of modern science is to discover the symmetry hidden within the tangle of ordinary life.

The universe began in an extremely hot state of utmost simplicity and symmetry. As it expanded and cooled, this perfect symmetry was broken, giving rise to the world of diversity and structure that we inhabit: galaxies, stars, planets, life.

To us today, the fundamental forces of nature appear distinct: gravity, electromagnetism, and the strong and weak nuclear forces. Gravity was conceived by Newton as a force exerted by objects in space, but Einstein reimagined it as the curvature, or "warping," of spacetime. Electricity and magnetism were understood as two "separate" forces until the 1870s, when James Clerk Maxwell proved that they were one. The strong nuclear force holds the nucleus together, while the weak nuclear force is responsible for the decay of certain elementary particles into other types.

The balance between these forces determines the existence and behavior of everything in the visible universe. Originally, it is theorized, all four forces were linked, and today scientists dream of finding a single set of equations describing all four. By colliding subatomic particles, physicists have discovered that at extremely high temperatures the differences between the forces begin to disappear. So far, theorists have succeeded in developing an "electroweak theory" that unifies electromagnetism and the weak force, and they assume that at a hot enough temperature, these two forces and the strong force would all be equal. More speculative theories seek to include gravity as well.

TIME TRAVEL TO THE BIG BANG

Imagine yourself journeying back in time—millions and billions of years—closer and closer to the moment of the big bang. The further you go, the hotter and denser the universe becomes, and broken symmetries are restored. As you approach the primal origin, the electroweak force and the strong nuclear force become identical: an electronuclear force. Finally, you arrive at 10^{-43} second after the universe came into being—a ten-millionth of a trillionth of a trillionth of a trillionth of a second. This moment is known as Planck time, named after the German physicist Max Planck. Earlier than this is hard to probe, because the density of matter becomes so great that the very nature of space and time is uncertain. At Planck time, presumably, gravity exerts as much strength as the electronuclear force. All interactions between elementary particles are indistinguishable. Perfect symmetry.

How did this symmetry of the beginning become so disguised over the course of time? Retrace your steps from Planck time toward the present. Before you even start, gravity has dropped out of the primordial unity. This soon in the early universe, things are still very hot: $10^{32\circ}$ Kelvin, or 100 million trillion trillion degrees, which is about five trillion trillion times as hot as the core of the sun. But the universe is expanding and starting to cool; its radiation and particles are losing energy. For an infinitesimal fraction of a second, there is still a symmetrical electronuclear force, but soon the forces become distinct in strength and range. At about 10^{-35} second after the beginning, when the temperature falls to 10^{27} degrees, the strong nuclear force decouples from the electroweak force. At about 10^{-11} second, the electromagnetic and weak forces branch apart.

Meanwhile, matter is also losing its oneness. By the time the universe is one billionth of a second old, there are four forces (gravity, electromagnetism, and the strong and weak nuclear forces) and about two dozen kinds of elementary particles. This fracturing of symmetry creates the particles of matter and energy found today around us—and within us.

Perfect symmetry may sound alluring, but it is sterile. If the primal force had not broken into four forces, the universe would be a very different place, if it existed at all. Tiny deviations from complete uniformity now give rise to nuclei, atoms, and molecules; then galaxies, stars, planets, and people. We exist today in our present condition, with all our flaws and imperfections, because of broken symmetry, just as Kabbalah teaches that our jumbled, blemished reality derives from the breaking of the vessels.

As a result of the breakage, it seems that we are no longer part of the One, and we act as if we are autonomous, exercising what *feels* like free will. Through *tsimtsum* (divine withdrawal) and *shevirah* (breaking), God has granted us a substantial measure of freedom, like a parent relinquishing control so that her child has room to grow. But every so often, we are captured by the transcendence of a sunset, enraptured by the immanence of a lover's touch. Handing a dollar to a needy stranger, we may be astonished by the warmth and gratitude in his eyes. Our sense of separateness is undermined by wonder or love, and our fragmentation is momentarily mended.

Mending, or *tiqqun*, begins with a change in perspective, a compassionate awareness. The jagged shards of material existence are seen anew, in light of their origin, before the shattering transition from infinity to boundedness, before the cosmic egg hatched.

The physicist peers back, too, to a time before the symmetry was broken. Or rather, to a time before it was "hidden," because symmetry lingers in the equations. Cosmology and Kabbalah share a Platonic perspective: What we observe is only an imperfect reflection of a deeper reality that displays symmetry or unity. Only because nature's symmetry is broken do the various elementary particles appear to have different properties. They are like facets of a cut diamond, shining distinctly as the diamond is turned in the light but, in fact, all manifestations of the same underlying object. If the temperature could be raised sufficiently, back to $10^{32\circ}$ Kelvin, the disguised symmetry would reemerge. Simply being aware of that fact enables the physicist, or anyone open to a scientific perspective, to perceive the world holistically.

THE PURPOSE OF IT ALL

Broken symmetry and the breaking of the vessels are distinct theories, each generated by a different approach to the question of the origin of the universe: Yet their resonance is intriguing. Timothy Ferris, an eloquent science writer, describes the search for symmetry as follows: "Physicists, in identifying the various elementary forces as having arisen from the breaking of a more symmetrical unified force, or in finding concealed symmetries cowering in the cramped nuclear precincts where the strong force does its work, are in effect piecing together the shattered potsherds of that perfect world."

Ferris did not intend to allude to Kabbalah, but his image betrays the resonance between the two types of shattering. The human mind has devised alternative strategies—scientific and spiritual—to search for our origin. The two are distinct but complementary. Science enables us to probe infinitesimal particles of matter and unimaginable depths of outer space, understanding each in light of the other, as we grope our way back toward the beginning. Spirituality guides us through inner space, challenging us to retrace our path to oneness and to live in the light of what we discover.

Oneness is not the whole story. There is value also in the unique individuality of each creature. The Hasidic mystics ask, "What delights God: oneness or individuality?" One answer we have already heard: "When we become aware that only God exists, then God receives from us the true delight He hopes for."

So it would seem that oneness is the goal. But there is a different, complementary answer: "God does not derive as much delight from us when we are all one entity as when we are individuals. For God desires us so that He can receive added delight from the detailed individuality within our separate bodies."

Each person expresses the oneness inimitably. What delights God is the variety and the immense spectrum of being. In the words of the founder of Hasidism, the Ba'al Shem Tov, "God wants to be served in all possible ways."

Having emerged from the cosmic pool of nothingness, each of us has woven a story of who we are. If we take ourselves and our tales too seriously, if we remain hermetically sealed within the borders of self, then we lose contact with the oneness of it all. If we become too attached to oneness, we forfeit the zesty spiciness of life and end up having sat out the entire dance. The middle path is being aware of the underlying oneness while expressing it creatively.

We are composed of energy, but it has congealed in each of us in a particular way. As the kabbalist Shim'on Lavi puts it, "With the concealment of the light, the things that exist were created in all their variety. This is the mystery of the act of Creation." Concealed oneness and hidden symmetry expand the possibilities of being. By manifesting energy in distinct patterns, they enable a universe to evolve.

But how can we find the symmetry hidden so deep in space-time? It disappeared fourteen billion years ago, fourteen billion light years away, in the first nanosecond. Yet its afterglow haunts our minds, as the big bang still echoes all around us through the cosmic background radiation. As the poet Paul Valéry wrote, "The universe is built on a plan, the profound symmetry of which is somehow present in the inner structure of our intellect."

From within, through meditation, each of us can glimpse oneness. One focus of spirituality is finding a balance between self and God, *bein adam la-maqom*, between me and oneness. This balance, at best, is impermanent: If I find it, I cannot hold it for long. But falling out of balance is also part of the balance: I am one with the cosmos—and, at the same time, I am an individual incarnation of the cosmos. With effort and a little luck, these two modes of being maintain a creative dialogue.

Mystics often use the image of a drop of water in the ocean to describe the relation between self and oneness. Realizing its true nature, the self dissolves in God. Perhaps another image is more meaningful for us today: the ocean and the wave. Momentarily, each wave seems to be a separate entity. But actually, no wave is distinct from the ocean, its ground of being. Waves are just surface manifestations of the ocean. Allowed its moment of apparent separateness by the forces of nature, each wave soon disappears back into the whole, as another takes its place. Similarly, each living being seeks to maintain its individual identity, though everything within the individual derives from its originating ground.

Once you dare to peek beneath the veil, to uncover the fabrication of self, you encounter possibility—and danger. It is terrifying to be set adrift in the vast ocean. Ethical dangers lurk, too; momentarily seeing through the self doesn't automatically put an end to selfishness. You discover that oneness is not enough, that spirituality doesn't work without finding the balance between self and other, *bein adam la-havero*. There is not just "me" and "oneness"; there are other human beings, other creatures, other incarnations of the oneness. Becoming aware of the "other" is the basis of morality. Look at the Ten Commandments, whose first word is *Anokhi*, "I" ("I am *YHVH* your God"). The first letter of the first word is *alef*, which is also the Hebrew number one. The Ten Commandments begin with oneness. But they end with the word *re'ekha*, "your neighbor": "You shall not covet anything belonging to your neighbor." The message is that just seeing another person, just encountering a human presence, should stimulate us to feel responsible toward him or her.

Rooted in oneness, the self branches out, seeking balance in both directions—back to oneness, out toward the "other." One can attain such balance by discovering oneness hidden in the other, by acknowledging the other as unique.

And finally, what is the goal of all this searching, of our cosmic game of hide-and-seek? This is how Abraham Abulafia puts it:

> The purpose of the marriage of a woman and a man is union.
> The purpose of union is fertilization.
> The purpose of fertilization is giving birth.
> The purpose of *that* is learning.
> The purpose of *that* is comprehending.
> The purpose of *that* is to maintain the endurance of the one who comprehends with the joy of comprehending.

The purpose of the game is to keep playing, to keep seeking, to keep discovering more about God, universe, self, other. The ecstasy of discovery enlivens the seeker. Anyone can play. In fact, we are all playing the game already. But to play well, you have to practice. Even before that, you have to learn the rules of the game, the rules of Torah.

Part Three

TORAH AND WISDOM

6. The Essence of Torah

All of Western religion emerges from the Torah. The rabbis expounded Torah, as did Jesus, though his apostle Paul believed that faith in Christ superseded Torah. The Koran describes Muhammad's revelation as an improved Scripture, closer to the heavenly original: "O people of the Book! Our Apostle has come to you, revealing to you many things of the Book that you have hidden and passing over many others."

The daughter religions of Judaism—Christianity and Islam—define themselves against the foil of Torah. But Torah is not a fixed entity, a stable background. It is more expansive than it appears—and more concentrated. By tracing the variations of Torah, we can discover how revelation unfolds, and how to return to the source, to experience revelation anew.

Torah is often translated as "Law," but this is inadequate. Torah implies teaching, showing the way. A guide through the welter of existence, Torah provides a compass and a path. Torah has three dimensions. What we usually think of as the Torah is the Written Torah, the Five Books of Moses from Genesis to Deuteronomy, the 304,805 letters from the *bet* of *Bereshit,* "In the beginning," to the *lamed* of *yisra'el,* the final letter of the final word at the end of Torah. But there is no end: *ein sof*. The Written Torah is expanded by the Oral Torah, an enormous mass of rabbinic interpretation recorded in the Talmud and Midrash.

The Oral Torah, as you can tell by its name, was not written down at first. It represents hundreds of years of clarifying and amplifying the Written Torah. In the words of the medieval poet Judah Halevi, "That which is plain in the Torah is obscure, all the more so that which is obscure." The terse style of the Torah, the vague generality of some of its commands, and the outright contradictions between certain verses made it impossible for life to be regulated solely in accordance with the Written Torah. Once again, Judah Halevi: "I would like to see someone try to adjudicate between two litigants on the basis of [the laws in Exodus and Deuteronomy]."

Many things are not spelled out in the Torah. To take one famous example, work is forbidden on the Sabbath, but we are not told exactly what "work" is. Over time, social and economic conditions changed; urban society developed in complex and unforeseen ways; new situations had to be addressed. The Torah itself is aware of the uncertainty of the future. In Deuteronomy, we read: "If a case is too baffling for you to decide, you shall appear before the levitical priests, or the magistrate in charge at the time. You shall act in accordance with the *torah* [the instruction] that they give you."

Like the universe, Torah expands continually and in all directions. As the Bible is applied to the here and now, its meaning unfolds. This process of interpretation and expansion is called *midrash*, which literally means "searching." Midrash is a way to search for and elicit the various meanings of Torah, meanings hidden in the text or sprouting from the fertile imagination of the reader. Each generation engages Torah and discovers a new aspect of revelation.

A good reader wrestles with the text of Torah, and neither reader nor text is ever the same. By encountering Torah directly, by questioning and probing its meaning, we rediscover and reimagine who we are, as we do by exploring the evolution of the cosmos, which is not separate from our evolution. Torah responds to our probing by evolving in countless unpredictable ways. The question "What does the text mean?" becomes more personal—"What does this mean for my life?"—and question turns into quest.

STANDING ON ONE FOOT

At times, overwhelmed by the sheer volume of Written and Oral Torah, the mind pauses and wonders: What is the essence? The impulse to expand Torah is balanced by the impulse to find its gist.

Tradition has fashioned an entire way of life—with specific do's and don'ts—based on the *mitzvot* of the Torah. According to Rabbi Simlai in the Talmud, there are 613 *mitzvot* in all: 365 "Thou shalt not's," corresponding to the number of days in the year; and 248 "Thou shalt's," one for every part of the human body. Each day is a struggle for self-control, each limb is holy.

There have been various attempts to formulate the essence of Torah, to identify the center of gravity around which spiritual life revolves. Rabbi Simlai himself, after noting that Moses received all 613 *mitzvot* from God, mentions various biblical heroes who reduced the number. King David expressed the essence in just a few verses of Psalm 15:

> *YHVH*, who may sojourn in Your tent, who may dwell on Your holy mountain?
> One who lives purely, who does what is right, and in his heart acknowledges the truth;
> whose tongue is not given to evil, who does no harm to his fellow,
> nor bears reproach for his acts toward his neighbor;
> for whom a contemptible man is abhorrent, but who honors those who revere *YHVH*;
> who stands by his oath even to his hurt;
> who does not lend money at interest, or accept a bribe against the innocent.
> One who acts thus will never stumble.

The prophet Isaiah was more succinct: "One who walks in righteousness, speaks uprightly, spurns profit from fraudulent dealings, waves away a bribe instead of grasping it, stops his ears against listening to infamy, shuts his eyes against looking at evil—such a one shall dwell in lofty security."

Isaiah's contemporary, Micah, managed to reduce all the *mitzvot* to three: "It has been told to you what is good, and what *YHVH* requires of you: only to do justice, love kindness, and walk humbly with your God." Not to be outdone, Isaiah came back and boiled it down to two principles: "Observe what is right and do what is just." Then came Amos and reduced them to one: "Seek Me and live."

One of the most famous attempts to put the whole Torah into a few words is Hillel's response to a potential convert. The gentile first approached Hillel's colleague and rival, Shammai, who was a builder by trade, and said, "Convert me, on condition that you teach me the entire Torah while I stand on one foot." Shammai drove him away with the measuring rod he was holding. Then, the gentile came to Hillel, who converted him, saying, "What is hateful to you, do not do to your fellow. This is the entire Torah; the rest, its commentary. Go and study."

Shammai refused to dilute the Torah. Hillel distilled its essence, but didn't stop there. First, he indicated that all the rest of the Torah is an attempt to spell out what its essence means. Then, he directed the convert to delve into the commentary.

Hillel's formulation is a negative version of the Golden Rule. Jesus, in the generation following Hillel, puts it positively: "Whatever you wish that people would do to you, do so to them. For this is the Torah and the prophets." Jesus' demand is more idealistic and extreme. Hillel's demand is more down-to-earth, more in the style of the rabbis. His "essence of Torah" is practical, while Jesus' essence is for the hasid, the impassioned devotee. But Jesus is simply paraphrasing the great *mitzvah* of Leviticus, in the middle of the Torah: "Love your neighbor as yourself, *ve-ahavta le-re'akha kamokha.*" Elsewhere in the Gospels, he teaches that this verse together with "You shall love *YHVH* your God with all your heart, with all your soul, and with all your might" constitute the two most important *mitzvot* of the Torah. Rabbi Akiva, who was born a decade or two after Jesus and, like him, was martyred by the Romans, also emphasized these two *mitzvot.* He refers to "Love your neighbor" as "a central principle in the Torah." As he was being tortured by the Romans, he celebrated the fact that now, finally, he could demonstrate his love for God with all his soul, "even if God takes your soul."

Love is the essence of Torah, the force that connects us with the oneness of the cosmos. In the words of an anonymous midrash, "Whatever you do, do it only out of love." Torah is a commentary on love, the ongoing Jewish attempt at spelling out how to love God and one another, how to get along with each other. At times, though, the commentary obscures the essence. Some of the *mitzvot* seem so anachronistic and inappropriate, even after all the ancient and modern efforts of Oral Torah. More basically, the very notion of "commandment" seems antithetical to the spontaneity of love. If we love God or our neighbor because we are responding to a command, what kind of love is that?

But without *mitzvot*, without specifics, how do we know how to act? We don't. And *that* is exactly the point. We cannot know, ahead of time, exactly what is required at this particular moment.

THE WISDOM OF THE PRESENT

In Egypt, Moses confronted Pharaoh with God's demand: "Let My people go and worship Me." When Pharaoh refused, God brought one plague after another. Finally, after nine plagues, Pharaoh was about to give in. He summoned Moses and told him that all of the Israelites—men, women, and children—could go to worship God, but their sheep and cattle must remain in Egypt, as a kind of collateral. Moses told Pharaoh that the livestock must come along, too: "Not a hoof shall remain behind, for we must select from them to serve *YHVH* our God. And we ourselves cannot know with what we will serve *YHVH* until we arrive there."

Until we find ourselves in a particular situation, we cannot know exactly how to respond. Tradition provides us with insights and guidelines, the accumulated wisdom of the past. We need access to these traditions; it would be foolish to overlook or discard them. But the past has no monopoly on wisdom; the present moment is what matters most, with all its potential and uncertainty.

The tradition itself recognizes this, even as it insists on observance of the *mitzvot*. A moment ago, we encountered Rabbi Akiva, a pivotal figure in the emergence of rabbinic Judaism. Akiva was tortured and killed by the Romans, supposedly for teaching Torah in public in defiance of a Roman edict. But Akiva's real crime may have been more political and messianic: He fervently supported Bar Kokhba (Shim'on Bar Koseva), the messianic pretender who incited a revolt against the Roman empire that took three and a half years to quell.

The Talmud relates that Akiva died with the words of the *Shema* on his lips: "Hear, O Israel: *YHVH* our God, *YHVH* is One." His dying words were repeated by countless Jewish martyrs from the Crusades to the Holocaust. Akiva was their role model: someone who persisted in believing in God despite what appeared to be a fatal act of divine abandonment. Akiva was happy to die for God because now he finally understood what it meant to love God "with all your soul." His tortuous death was redeemed and uplifted by the discovery of a new way to fulfill a *mitzvah*. As his breath gave out, he held on to the last word of the first line of the *Shema* as long as he could: "*Ehaaaaa-d*," so that "his soul expired in *one*." Akiva died into oneness.

Often ignored is why Akiva chose to pray the *Shema* at that moment. Was he looking for the most profound verse he could find to serve as his passionate farewell to his students? An appealing interpretation, perhaps, but that is not what the Talmud says. Rather, Akiva prayed this prayer simply because "it was time to recite the *Shema*," *zeman qeri'at shema hayah*. His heartfelt declaration resounded through history, but Akiva was just doing what a religious Jew does at sunset or sunrise: pray the *Shema*. He wasn't going to be thwarted by some Roman executioner.

One of Akiva's themes is "the suffering of love," *yissurin shel ahavah*, experienced in one's love for God, or God's love for him. Akiva's death is the ultimate enactment of that teaching. But in his death, Akiva teaches us something else as well: doing what needs to be done at this particular moment.

But aren't certain *mitzvot* more significant than others and certain times more holy? In the sixteenth century, Rabbi David ibn Zimra was asked about this, concerning a Jew who found himself in jail: "Reuben was imprisoned. He could not go out to pray in a group of ten or to fulfill the *mitzvot*. He pleaded before the officer in charge, who refused to let him go, except for one day a year, whichever day he wished. Let the teacher teach: Which day among all the days of the year should Reuben select to go to the synagogue?"

In his response, Ibn Zimra cited an authority who determined that Yom Kippur is the appropriate first choice, but he disagreed. "The first *mitzvah* that is available to him—and which it is impossible to fulfill because of his imprisonment—takes precedence. He should pay no attention to whether the *mitzvah* that encounters him first seems important or unimportant, for you do not know the value of the *mitzvot*. To me, this is very simple."

The holiest moment is *now*. The holiest act is whatever you can do right here and now in our little corner of the cosmos. In Hasidism, this approach is developed further and applied enthusiastically: The fixed menu of *mitzvot* is expanded. As Dov Baer said, "One who fulfills 'Know God in all your ways' has an infinite number of *mitzvot*." No book can record all the ways or prescribe all the *mitzvot*, all the opportunities to contact and experience God. In the words of the *Ba'al Shem Tov*, "When you are walking and talking with other people and cannot study Torah, maintain your union with God by contemplating oneness. Similarly when you are on the road and cannot pray or study as you usually do, serve God in other ways, and do not be distressed about this because God wants to be served in all possible ways."

The essence of Torah cannot be captured in words or commands. It is glimpsed unexpectedly. One who claims mastery of the essence misses the whole point: The Torah is an unending, unpredictable flow. The *Zohar* conveys a similar message in a parable about a mountain man.

> There was a man who lived in the mountains. He knew nothing about those who lived in the city. He sowed wheat and ate the kernels raw.
>
> One day he entered the city. They brought him good bread. He said, "What is this for?" They said, "Bread, to eat!" He ate, and it tasted very good. He said, "What is it made of?" They said, "Wheat."
>
> Later, they brought him cakes kneaded in oil. He tasted them and said, "What are these made of?" They said, "Wheat."
>
> Finally, they brought him royal pastry made with honey and oil. He said, "And what are these made of?" They said, "Wheat." He said, "I am the master of all of these, for I eat the essence of all of these: wheat!"
>
> Because of that view, he knew nothing of the delights of the world; they were lost to him. So it is with one who

> grasps the principle and does not know all those delectable delights deriving, diverging from that principle.

The mountain man has the essence: the raw words of Torah. But he is wallowing in essence—reducing everything to what he already knows—so he cannot appreciate the ways in which the raw material of Torah is transformed into midrash, allegory, and mysticism. Torah should be experienced in all its flavors, just as life should be lived in all its variety. These are the delights deriving and diverging from the essence.

According to the uncertainty principle, as we have seen, it is impossible to simultaneously determine both the location and velocity of subatomic particles. The particles cannot be pinned down; they behave like waves. A similar principle applies to Torah: You may think you know precisely what a word means, but who knows how it will unfold and where it might take you? Even as a particular meaning is established, waves of interpretation begin to undulate. Meanings are conveyed back and forth between word and reader. Just as the observer of a subatomic particle constructs reality through the very act of observing, so the meaning of Torah is continually reinvented in the reader's creative imagination. No matter how hard you try, you cannot observe the text "as it is." It is *not*, until the sequence of letters evokes a meaning in the mind—a meaning that is instantaneously reflected back to the letters, transforming them into *devarim*: things, words. The mind constructs both things and words. The essence is that there is no single essence.

7. The Ripening of Torah

There is another dimension of Torah, the source of both the Written Torah (the Five Books of Moses) and the Oral Torah (their ongoing interpretation). Kabbalists call this *Torah Qedumah*: Ancient Torah, Primordial Torah, the dimension of Torah that is timeless. Primordial Torah is Divine Wisdom (*Hokhmah*), the first ray to emerge from the creative nothingness of *Ayin*. As with sunlight, it has no color. It has yet to be refracted through the prism of language into the spectrum of meaning. Such wisdom cannot be spelled out in words. Primordial Torah cannot be reduced to writing; it cannot be mouthed or transmitted directly—but it is the source of all Torah, Written and Oral. According to Chinese philosophy, "The *tao* [the "way"] that can be named is not the eternal *Tao*." A kabbalistic paraphrase would be: "The Torah that can be named is not Primordial Torah."

The rabbis of the Talmud do not mention *Torah Qedumah* by name, but they describe something similar. Torah, they say, precedes creation. God gazed into the Torah and created the world, just as an architect follows blueprints. One could understand this in a simple-minded way: God rolled open the Torah scroll to Genesis, chapter 1, saw the verse "God said, 'Let there be light,' and there was light," and knew exactly what to do on day one. But the implication of the midrash is more profound: There is a dimension of Torah that encompasses the divine mind. By contemplating Primordial Torah—primordial cosmic law—God imagined a model of the universe. This harmonizes with Einstein's view that the laws of nature reflect how God thinks.

What is the relation between *Torah Qedumah* and the Torah that we now possess, the Torah that possesses us? In another midrash—remarkable and rarely cited—we read that Torah is "an unripe fruit of heavenly wisdom." If something is unripe, it's incomplete, as in the other examples provided by the same midrash: "Sleep is an unripe fruit of death. A dream is an unripe fruit of prophecy. Shabbat is an unripe fruit of the world-to-come. Torah is an unripe fruit of heavenly wisdom."

What is unripe provides only a fraction of what is fully grown, as in the Talmudic variation: "Fire is one-sixtieth of hell; honey, one-sixtieth of the [sweetness of the] manna; Shabbat, one-sixtieth of the world-to-come; sleep, one-sixtieth of death; a dream, one-sixtieth of prophecy."

In what sense is the Torah "unripe"? As Moses de León said, "But isn't the Torah the source of life? How can you say that she is unripe? What is unripe is inferior! Like fruit that falls prematurely from a tree and ripens on the ground: It isn't as good as fruit that ripens on the branch."

The midrashic metaphor is shocking because, traditionally, Torah represents the will of God as revealed at Sinai to Moses and Israel, who accepted it for eternity. What could be more perfect? But the Torah is "unripe" because the particular form it has assumed—the Written Torah—is just one possible reading of Divine Wisdom, of Primordial Torah. The Written Torah emerges from Primordial Torah. Traditionalists believe that God revealed the Written Torah to Moses at Sinai beginning on the sixth day of the month of Sivan during the last third of the thirteenth century before the common era. Modern biblical scholars contend that the Torah was spliced together from various sources by several editors hundreds of years later. In either case, the Torah as we know it is not identical with Primordial Torah. It is a pale reflection, an unripe fruit.

HOW TORAH EXPANDS

Torah as we read it is not the whole story. In fact, there is not even one absolutely correct way to read it. Speculating on why the Torah is written without vowels, the thirteenth-century kabbalist Bahya ben Asher of Saragossa, wrote that "without vowels, the consonants bear many meanings and splinter into sparks. Once vowelized, a word means just one thing. Without vowels, you can understand it in countless, wondrous ways."

The Torah must remain vowelless to preserve its open-ended nature and to ensure that the words continue sparkling with new, unforeseen meaning. In the words of another thirteenth-century kabbalist, Jacob ben Sheshet of Gerona: "The Torah scroll may not be vowelized—so that we can interpret every single word according to every possible reading."

In chapter 1, we touched on the theory that our universe is just one of countless universes, each of which differs in its initial conditions and its types of matter, maybe even in its laws of physics. Similarly, according to one kabbalistic theory, our universe is just one of many, each of which is governed by its own set of laws: its own version of Torah. Each universe emerges out of one of the seven lower *sefirot* in a cosmic cycle lasting seven thousand years and characterized by the particular qualities of that *sefirah*. In each cycle, the Primordial Torah is read differently, its letters arranged into words appropriate for that cycle. Our current version of the Torah, with all of its do's and don'ts, corresponds to the *sefirah* of *Gevurah*—sternness, rigor, judgment. This Torah suits our conflicted and unfulfilled state of being in which we are caught between good and evil. In the preceding cycle, which ended about 5,800 years ago, the *sefirah* of *Hesed*—grace, love—set the tone: There was no evil impulse and no need for restrictions and prohibitions. Then, the Torah was revealed according to *Hesed*. In the coming age, the Torah will reflect the *sefirah* of *Tif'eret,* the harmony of these opposite spheres.

Like life and the universe, Torah evolves. The current version of the laws of Torah is not eternal and applies to a limited domain. Similarly, the history of science reveals that the laws of nature apply to a limited domain—their domain of validity. For example, Newton's laws of physics were accepted because they accorded with experiment. When experiments became more accurate, this set of laws turned out to work well only in the domain of speeds that are small compared to the speed of light. Physicists struggled to understand what was going on at higher speeds, at the boundary of the domain of validity. In formulating special relativity, Einstein discovered a new set of laws that were valid inside, near, and beyond the boundary—at low speeds and at speeds approaching the speed of light. But special relativity does not include the effects of gravity. That domain required another set of laws: Einstein's general relativity. Near the singularity inside a black hole, these laws also fail, to be replaced by the laws of quantum gravity.

Perhaps there is one ultimate set of the laws of nature for our universe, but the current version of these laws is an approximation. Similarly, our current version of Torah is an approximation of Primordial Torah: an unripe fruit.

How can Torah ripen for us today? How can we experience its full taste? First, by listening. If a word of Torah speaks to us right now, it ripens instantly. For example, the first question posed in the Bible is a single word, directed by God to Adam: "*Ayekah*," "Where are you?" If we pause and listen, the question becomes real, reverberating inside each of us as we wander through life, often hiding from what we know or what we fear: Where am I? What am I doing with my life?

Torah can be heard right now as if for the first time. Describing the arrival of the Israelites at Mount Sinai, Exodus states: "On the third new moon after the Israelites left the land of Egypt, on this day they came to the Wilderness of Sinai." Wondering why it is written "on *this* day" and not "on *that* day," the Midrash explains, "Every day that you study Torah, say, 'It is as if I received it this very day from Sinai.'"

Sometimes, though, listening doesn't do the trick. We listen and hear something objectionable: animal sacrifices, for example, or "an eye for an eye," or a patriarchal lifestyle. Can these notions ripen or should they just be discarded? In such cases, ripening comes about through historical change and evolution, through the reformulations of Oral Torah, including our own.

After the Temple was destroyed in 70 C.E., sacrifices gave way to more spiritual forms of worship: prayer, study, and meditation. Sin could no longer be atoned for by sacrifices, but instead, in the words of the first-century Rabbi Yohanan ben Zakkai, "we have deeds of kindness." The rabbis softened and humanized the primitive justice of the Torah, interpreting away the literal sense of "an eye for an eye" and replacing it with the principle of monetary compensation. If I gouge out someone's eye, my eye isn't gouged out in return. Rather, says the Talmud, I have to pay compensation for loss of vision, pain, medical expenses, unemployment, and the humiliation of being disfigured.

Patriarchy persists in religion and society, but feminists have taught us to question our attitudes, behavior, and language. If we are serious about confronting sexism, we have to change the ways we conceive and talk about God. To quote the feminist Mary Daly, "As long as God is male, the male is God." The ripening of Torah requires us to realize how far we've come and how far we still have to go.

At times, we listen intently—and the Torah ripens. Other times, it doesn't taste right, but it can ripen through a midrash of the rabbis or from our own brand of midrashic creativity. Torah evolves, as our own situation evokes new meaning. It is we who help Torah expand. In the words of the kabbalist Jacob ben Sheshet, "It is a *mitzvah* for every wise person to innovate in the Torah according to his capacity." Imaginative creativity resembles revelation. As Ben Sheshet explains to his readers, following one of his insights: "Do not think that this is far-fetched. If I had not invented it in my mind, I would say that it was transmitted to Moses at Sinai."

THE INFINITIES OF TORAH

What did God actually reveal at Sinai? One Talmudic source insists that the entire Written Torah was given to Moses right then and there. Another suggests that Moses received the revelation "scroll by scroll," i.e., section by section, as the Israelites wandered through the desert. But what, according to tradition, did God actually say publicly to all the people? The Ten Commandments? Well, look at them closely. Only the first two are written in the first person: "I am *YHVH* your God who brought you out of the land of Egypt, out of the house of bondage. Do not have any other gods before Me." The remaining eight commands are in the second person: You shall do this; you shall not do that. Maybe, as another rabbi suggests, just the first two commands were spoken directly by God: "from the mouth of Power," *mi-pi ha-gevurah*. Even this bit of revelation overwhelmed the Israelites. Terrified, they asked Moses to intercede and convey to them the rest of the message, to serve as a vessel for the surging divine flow.

Maybe God spoke only the first command, or only the first word: *Anokhi*, "I am." There is a later, Hasidic view that goes even further: Only the very first letter of the Ten Commandments was spoken by God. And what is the first letter? The *alef* of *Anokhi*. An *alef* without a vowel has no sound. It simply represents a glottal stop, a position taken by the larynx in preparation for speech.

A silent *alef* indicates the beyond. It frames a window onto the infinite realm of possibility. A kabbalist would appreciate the German mathematician Georg Cantor's decision to symbolize various types of infinity with the *alef*. The *alef* is the uncarved block, preceding the shaping of words, the verbal formulations of Torah. It represents pure potential, with no specific content spelled out. The Written Torah, the Five Books of Moses, is already a commentary on the *alef*, a series of glosses inscribed in its boundless margins.

The silent, pregnant *alef* resonates with another mystical insight. The Talmud teaches that the Torah was originally written with "black fire on white fire." According to one kabbalist, this implies that the white space behind the letters is the essential Torah. The letters and words represent merely one interpretation of the white expanse.

One could conclude that all the documents of tradition betray the pristine nature of the *alef*, that no words can convey its vastness. Alternatively, one could say that the *alef* includes the entire tradition. In a sense, as the Midrash insists, everything was already given at Sinai. "That which the prophets in every generation were destined to prophesy, they received at Sinai. Even the wise who appear in every generation—each one received his wisdom at Sinai. Even what a student asks his teacher, the blessed Holy One said to Moses at that time."

The *alef* of revelation cannot be pronounced, yet it finds expression moment by moment. Its ineffable oneness can be rendered only in fragments, only by being translated into the duality of language, where each word symbolizes a separate thing. Revelation turns into Torah—into this and that, do's and don'ts, countless questions and answers across the generations. The *alef* generates Torah, which characteristically begins with the *bet* of *Bereshit* (In the beginning), letter number two. In the words of the Psalms, "One, God has spoken; two have I heard." *Ahat dibber elohim, shetayim zu shama'ti*. Similarly, the multiplicity we experience in the world derives from a primordial unity fractured by the breaking of the vessels, a symmetry broken in the big bang.

The *alef* is the essence that cannot be grasped, an indication that essence cannot be packaged. The Written Torah—written in the margins of the *alef*, written to interpret essence—has been mistaken for essence. Accustomed to this interpretation, can we appreciate the raw power of Primordial Torah? One particular reading of the divine mind has guided us for millennia and still has a hold on us. Through midrash, we have stretched the Written Torah, applied and contemporized it, personalized it; we have, in effect, interpreted the interpretation. But can we return to the source of it all? Can we uncover the unadorned *alef* and perceive the timbre of revelation: the sound of sheer silence, *qol demamah daqah*?

In other words, can revelation happen again? In considering this question, we should consider the inherent tension between religion and revelation. Religion is based on revelation. Revelation leads inevitably to tradition, but new revelations threaten tradition. By nature, religion is conservative; and revelation, revolutionary. The structures of Judaism and of any religion are built on the assumption that the canon is closed, that prophecy has ceased. With rare exceptions, revelation is relegated to the past, or to the Messianic future. In the words of a midrash, "The Torah that one studies in this world is nothing compared to the Torah of the Messiah." A new revelation would be an eruption. Who knows what it could do—or undo?

LISTENING TO THE ALEF

Yet, once we realize Torah's relativity, the *alef* beckons. New revelations are dangerous, but if we proceed with care for ourselves and others, and with respect for tradition's positive legacy, we can venture into the white space with a reasonable hope of emerging in peace, in one piece.

Though tradition guards against new revelation, its own description of what happened at Sinai can guide us toward the *alef*. According to one midrash, "When the blessed Holy One gave the Torah, no bird chirped, no winged creature flew, no cow mooed. The *ophanim* [angels shaped like wheels] did not fly; the *seraphim* [fiery angels] did not chant, 'Holy, holy, holy.' The ocean was still; human beings did not speak. The entire world was wrapped in silence. Then the voice came forth: 'I am *YHVH* your God.'"

Silence is essential. We cannot hear something new if we are babbling away or carrying on a constant internal monologue. By quieting the mind, by emptying ourselves of preconceptions, we create a space for the *alef* to spell itself out in new ways. Why was the Torah given in the desert? Because "one cannot acquire wisdom or Torah unless one makes oneself like an ownerless desert."

To enter the desert—to *become* the desert—we must surrender our images. The vast wilderness mocks our mental habits, and its stark grandeur inspires the "I" to renounce the claim to who it thinks it is, to see through itself—if only for a moment.

Even if the self becomes transparent, the next moment of revelation is mediated to each person differently. The message is not standardized, but designed for the individual. As Rabbi Yose son of Hanina said, "The divine word spoke with each person according to his capacity." Rabbi Levi offered this image: "The blessed Holy One appeared to them [at Sinai] like a statue whose face is seen from many directions. A thousand people gaze at it, and the statue looks back at each one of them. Thus, when the blessed Holy One spoke, each and every Israelite said, 'The word is speaking with me.'"

Revelation comes not only from on high, beyond, and above, but from within. Hundreds of years before Sinai, according to the Midrash, Abraham experienced revelation through his inner being, before Torah had been fixed in stone or captured in the Book. Abraham is a model for trusting our inner access to wisdom. It may seem presumptuous to imitate Abraham, founder of the faith, but the Midrash encourages such imitation: "Every single person should say, 'When will my deeds approach those of Abraham, Isaac, and Jacob?'"

The *alef* is here, right now, simultaneously beyond and within us. We can contact it, but are we ready for it? Can we trust how it will unfold in our minds and hearts? Can we trust ourselves? How can we evaluate our stammering, contending inner voices? Perhaps God speaks to us through our conscience, but still, even when we know or sense what is right, we often fail to carry through.

The *alef* is the source of everything verbal and mental, of all language and thought. Does that mean everything we think and say is a form of revelation? If God is the oneness of it all, does that make everything divine? In the words of a pantheistic Hasidic rabbi, "Know that all is God. Your thought and speech and all your vitality is divinity. This is the meaning of the verse in Psalms: 'Do not have a strange god among you.' This means: 'Do not think it strange that the divine is within you.' Know that all is divine."

The original context of the biblical verse is idolatry, but here the words have been transformed into a mystical proof text. Now, the message is: Cultivate an appreciation of the divine spark within you. Fan the spark into a flame.

It is one thing to say that divine inspiration manifests itself in our conscience, in flashes of insight or profound ideas. But in every thought, every fantasy? This sounds as absurd as it is dangerous. Yet, if God pervades everything, then our mental energy itself is a divine phenomenon, as is all energy and all the frozen energy that we call matter. The physical universe is God's body, and revelation is the radical awareness of this mystical fact.

But what will we do with this awareness?

We may realize that the energy coursing through us is the same energy that pervades all life, but we often think, act, and speak in ways that are banal, if not worse. Revelation provides awareness, not a detailed plan. To listen to the *alef* is to pause and intuit our connection to something beyond self. We do not know where the *alef* will lead us. Perhaps our next thought will be selfish, depressing, or violent. The *alef* is open-ended, and we are capricious, so uncertainty dominates. Yet, as we remind ourselves of our link to everything that exists, we challenge ourselves to channel the energy more effectively and harmoniously.

Having encountered the *alef*, we see Torah in a new light. No text can render the experience of revelation, but Torah focuses on what comes *after* revelation: how to live when we come down from the mountain. We need ethical guidelines, and Torah is a never-ending experiment in formulating them. Having exposed ourselves to the *alef*, we can now read some of the white space behind the letters of the Torah. Some of what we know from the Written Torah will need revision. Guidelines are not absolute. They change in the light of other wisdom, other renderings of the *alef*, such as philosophy, ethics, and the insights of other religions of the world. "Who is wise?" asks the Mishnah. "One who learns from every human being."

We need to find a balance between tradition and discovery; each is incomplete without the other. The *alef* can leave us stranded in infinity, while the Torah of the past can constrict and bind.

The ripening of Torah is ongoing. In the Bible, Isaiah hears God announcing a new revelation: "A Torah will come forth from Me." Jeremiah, too, hears God describe a new covenant: "I will place My Torah in their inmost being and write it upon their hearts." This new-ancient Torah, this Torah of the heart, is not reserved for some far-off messianic date. Constantly unfolding and as dynamic as the cosmos, it will never be completely spelled out, written on paper or parchment. It's too alive. Rooted in the soil of tradition, nourished by the endless *alef*, this tree of life branches into compassion and justice. The ramifications of these branches guide us through life, yielding fruit ripened by wisdom.

8. *Halakhah*: Walking the Path

The Torah enumerates specific *mitzvot* to guide how we act. Through Oral Torah, rabbinic Judaism has fashioned a code of law and religious practice, an entire spiritual discipline: *halakhah*. But if God is the oneness of it all, who is commanding me to follow this code?

Should I follow *halakhah* for the sake of tradition, to link myself with generations of my people? This approach has some validity. After all, I didn't pop up out of nowhere. I proceed from a rich past, which I honor by integrating it into my life. This argument is particularly compelling after Auschwitz, when maintaining the continuity of the Jewish people is itself a *mitzvah*, what philosopher Emil Fackenheim calls the 614th commandment.

If the God of tradition died at Auschwitz, then He had been terminally ill for some time. Yet for some post-Holocaust Jews, tradition lives on, unchanged. Through their intense religious commitment, they deny the absence of that God. It is as if they are saying, "Nothing has really changed. Since I am still obeying God's commands, He must be here." Questioning patterns of belief and behavior is hard, so it sometimes seems easier to keep doing the same thing.

We should take tradition seriously enough to challenge it, wrestle with it, help it evolve. We should not follow it slavishly, but neither should we oppose it dogmatically. Dogmatic opposition is just another type of enslavement, an indiscriminate reflex action. We should study and contemplate tradition, while at the same time being open to new expressions of wisdom, to fruit that is ripe for us.

Halakhah literally means "going" or "walking." *Halakhah* is intended to show us how to apply Torah day by day, moment by moment. Originally an oral tradition, *halakhah* became codified only gradually, first in the Mishnah (200 C.E.) and later in the Gemara (500 C.E.) and a series of medieval codes. But when *halakhah* becomes too fixed and rigid, behavior becomes routinized and we forfeit spontaneity. *Halakhah* succeeds through flexibility, which is perhaps why it was originally forbidden to record *halakhot* (specific halakhic decisions). In the words of Rabbi Yohanan bar Nappaha, whose teachings comprise a major portion of the Jerusalem Talmud: "One who writes down *halakhot* is as one who burns the Torah."

Completely spelling out a teaching ruins it. Without room for alternative expressions, *halakhah* is stultified, and religion is transmogrified into obsession or spiritual masochism.

SIMPLE GUIDELINES

The God I believe in doesn't provide a complete program for how to proceed through life, but I treasure the gems of Jewish tradition and the insights of those who have walked and wondered before me. By testing these as I walk along the way, I begin to fashion a flexible *halakhah*. Unlike the current orthodox code of law, which insists on precision and finality, my *halakhah* has more questions than answers. I stumble, and I learn from my stumbling. Or, as a Chinese Buddhist teacher said, "My stupidity is my bread and butter."

To start, I need a few simple guidelines, something to remember as I walk out the door in the morning. "Love your neighbor as yourself." Or Micah's triad: "Do justice, love kindness, and walk humbly with your God." But, as we have seen, the essence of Torah is that there is no essence. The challenge is to flesh out abstract principles, to actualize them in countless ways as we are confronted with each new challenge and opportunity. Still, it helps to keep verses like these in mind—and in heart. In sifting through the mass of Written and Oral Torah and traditional *halakhah* to determine what works best for us today, the prime criteria are love of God and neighbor, justice, and kindness. Encountering each particular *mitzvah*, I ask myself: Does this *mitzvah* enable me to express love toward my fellow human being? Will it train me in compassion? Will it contribute to a more caring community, a more just society, a sustainable planet? Can it enhance my spirituality and that of others? Such questions set the parameters of my *halakhah*.

Mitzvot fall into two categories: those between one human being and another (*bein adam la-havero*) and those between each human being and the oneness of God (*bein adam la-maqom*).

Most of the interpersonal *mitzvot* and much of the traditional *halakhah* based on them represent a code of justice and compassion. Life in the world exposes me again and again to the "other," to thousands of distinct human incarnations of the oneness. The sheer presence of the "other" calls out to me. I often avert my eyes, plug my ears, harden my heart to avoid it. The interpersonal *mitzvot* remind me of my responsibility—my capacity to respond—to the needs of others. For example, the book of Exodus commands, "Do not oppress a stranger, for you know the *nefesh* of a stranger, having yourselves been strangers in the land of Egypt."

Nefesh is often translated as "soul," but here the sense is that "you know what it feels like to be a stranger." Since you went through the experience of slavery, of being mistreated and degraded, you know what this is like. Now that you have been liberated, you should care for those who are still enslaved, for those who feel out of place, for the homeless, for the strangers, for those who *seem* strange. This *mitzvah* not only resonates with "Love your neighbor," but outdoes it by expanding it to include the stranger. This natural expansion appears explicitly in the next book of the Torah, Leviticus: "When a stranger resides with you in your land, do not wrong him. This stranger living with you shall be like one of your citizens. Love him as yourself, for you were strangers in the land of Egypt. I am *YHVH* your God."

Many interpersonal *mitzvot* express justice and love: Establish courts of law and fair legal procedures. Return lost property. Be ethical in business dealings. Honor your parents. Help the poor (e.g., by leaving corners of the field unharvested and by providing interest-free loans). Do not steal, murder, commit adultery or incest, bear false witness, return a runaway slave, covet, slander, bear a grudge, or be vengeful.

Naturally, my *halakhah* includes these do's and don'ts. But an example of an interpersonal *mitzvah* that doesn't fit into my *halakhah* is this admonition from Leviticus: "If a man lies with a male as one lies with a woman, the two of them have done an abhorrent thing; they shall be put to death." The Torah's own principles of love and compassion, and our contemporary understanding of the development of sexual identity, lead me to renounce this condemnation of homosexuality. As we have seen, the rabbis of the Talmud themselves transformed other harsh *mitzvot*, such as "an eye for an eye, a tooth for a tooth," which they reinterpreted to refer to monetary compensation. From their perspective, this was not a compassionate reinterpretation, but the "real" meaning of the text. Our historical consciousness enables—or perhaps compels—us to separate the Torah's intention from our evaluation.

My *halakhah* requires one general corrective to the interpersonal *mitzvot*. Those that are limited to fellow Jews, either by the Torah itself or in later tradition, should be broadened to include all humanity. Who *is* the "neighbor" in "Love your neighbor as yourself"? Who *is* "your brother" in numerous *mitzvot*? Despite whatever historical, sociological, or psychological circumstances contributed to such original limitations, we now need a more global vision to accord with the fact that each of us belongs to an emerging world culture. Big bang cosmology is important to our culture because it provides an account of our origins. Being aware that we have all sprouted from a primordial seed has profound ethical implications: Essentially, we are not that different from each other and our actions should reflect this.

The other traditional category of *mitzvot* is between each of us and God, though this distinction is artificial since God is often discovered interpersonally. Many *mitzvot* in this category do not apply today since they pertain to the Temple in Jerusalem: the sacrificial cult, the duties of the priests and Levites, offering tithes and the first fruits of the season. Other *mitzvot* of this type forbid various forms of idolatry and divination. How does my *halakhah* respond to some of the major *mitzvot* "between us and God," such as Shabbat, *kashrut*, and prayer?

HOW PRAYER EVOLVES

Let's start with prayer. After the Temple's destruction, worship changed radically. The sacrificial cult came to an end, and the synagogue, an institution already several hundred years old, became more vital. Prayer became decentralized, and rabbinic Judaism gradually elaborated a standard liturgy for the synagogue, which replaced sacrificial worship. In the words of one Talmudic rabbi, "The prayers were instituted to replace the daily offerings." But a contending statement traced this radical innovation to an ancient source—the Patriarchs. "Abraham instituted the morning prayer; Isaac, the afternoon prayer; and Jacob, the evening prayer." A new tradition was born—and instantly given an ancient pedigree.

For nearly two millennia, the *siddur* (the traditional "order" of the liturgy) has been a unifying factor in Jewish culture. While constituting our inherited archetype of Jewish prayer, it needs to be brought into harmony with our current understanding of God and the cosmos.

The American synagogue is now being transformed. One aspect of this transformation is the feminist reworking of liturgy, which balances the masculine language of the prayer book with feminine language or with gender-neutral prayers. Contemporary poet Marcia Falk has refashioned numerous prayers and blessings. For example, the *motsi*, the blessing over breaking bread, which traditionally reads: "Blessed are You, *YHVH* our God, King of the world, who brings forth bread from the earth." Falk's *Hamotzi'ah* is:

Let us bless the source of life
that brings forth bread from the earth.

Her "Blessing after the Meal" is a succinct, powerful alternative to the long traditional version.

Let us acknowledge the source of life,
source of all nourishment.

May we protect the bountiful earth
that it may continue to sustain us,

and let us seek sustenance
for all who dwell in the world.

The *Amidah*, the silent standing prayer, opens with a line that recalls the ancestors of Israel. Falk's *Amidah* begins:

Recalling the generations,
we weave our lives
into the tradition....

As we bless the source of life
so we are blessed.

Falk transforms a blessing to God "who restores His *Shekhinah* to Zion" into:

> *Let us restore Shekhinah to her place*
> *in Israel and throughout the world,*
> *and let us infuse all places*
> *with her presence....*
>
> *As we bless the source of life*
> *so we are blessed.*

As we continue to conceive of God in new ways, our mode of prayer will naturally change. The Holocaust is no less a challenge to our style of worship than was the destruction of the Temple. Then, the sacrificial cult was eliminated. Now, with our traditional concept of God undermined, to whom can we address our prayers? The Holocaust was far more efficient and deadly than the Roman assault on Jerusalem, and its theological impact is more acute and potentially devastating.

Feminism, the Holocaust, and the big bang all affect how we pray. The discoveries of contemporary cosmology are as wondrous as anything ever imagined, if not more. For the first time ever, humans are not simply imagining the beginning: We are *observing* it, in the cosmic background radiation, which has been traveling to us since the universe became transparent to light 400,000 years after the big bang. Cosmologists, whose specialty has always been the science closest to theology, are today piecing data together into "humanity's first verifiable creation story." This emerging cosmology stimulates us to develop a cosmic perspective.

Knowing the little that we do about how the world expanded into existence, our sense of awe is invigorated. We can imagine a new meaning behind the traditional name of God chanted in the morning service: "The One who spoke and the world came into being." The divine language is energy; divine grammar, the laws of nature.

But the fact that we know something about how the world began cannot help us address various other prayers. Knowing how the world ended for six million Jews, how can we simply repeat lines of prayer praising the God of history: "Righteous is God in all His ways, gracious in all His acts .... guarding His people Israel forever"?

If the traditional image of God is basically flawed, how can I pray with the same old words? If God is the energy of the universe, manifesting here and there as matter, what kind of prayer is appropriate or possible? I feel gratitude for the gift of life, but whom can I thank? I marvel at the green of this leaf, at the rustling wind, at the chirping of an unseen bird. My life is bound up with the life of all things. But what does this have to do with praying in the synagogue?

I am not suggesting that we discard the prayer book. Problematic as it is, our liturgy still resonates, enriched by centuries of chanting and devotion. Issues of gender and post-Holocaust theology have stimulated new prayers, as we have seen. Here, I want to propose something more modest, though in a way more radical.

SILENT PRAYER, BREATHING PRAYER

We need more silence in prayer. The words have been piling up for millennia; there are simply too many of them. This tendency toward wordiness reaches a crescendo on the High Holy Days, when Jews throughout the world read a thick volume of prayers over three days. Half the book is read on Yom Kippur alone. Words can be powerful, but an excess of them burdens the soul. The traditional Sabbath service is so long that to complete it in less than two and a half hours, the text must be read or scanned so quickly that anyone praying can easily suffer spiritual indigestion. The same applies to the daily morning service, which is briefer, consigned to about forty minutes. This service often turns into a communal race, punctuated by rapid page-flipping, so that the ten or so loyal congregants, who woke up half an hour earlier than they might have otherwise in order to go to *shul*, can still get to work on time.

We need fewer words—and more room for silence. We need fewer prayers—and more time for reflection. The prayers selected should be prayed slowly; a single verse can be chanted again and again. The best lines of prayer lend themselves to meditation.

If we limit ourselves to spoken words, we never taste the richness of silence. Incessant verbal prayer, with its duality of "we" and "You," maintains a mental barrier between us and God. In silence, we can contemplate the possibility that the barrier is an illusion.

Such epiphanies are rare since the self so infrequently opens the window to oneness. Undeniably, especially in our culture, we are immersed in language and in the mental language of thought. So we need words of prayer to express our pain and our joy, our longings, fears, and hopes. With words, we can cry and sing and bridge the divide between God and us. Even while crossing the bridge and sensing that we are part of what seemed "other," the part can still sing to the whole, addressing it personally and poetically as "You," although ultimately, we are not separate from You. Consider, for example, these lines selected from different parts of the morning service: "My God, the soul You have breathed into me is pure." "In Your light we see light." "With Your Torah, open my heart."

Precious words of prayer should not be rushed into oblivion. Allowed to linger, they lead to meditation. You can move back and forth between contemplating the personality of God and Its infinity. Meditative prayer overcomes the apparent dichotomy between the two, enabling us to develop intimacy with what seemed beyond.

The *Qedushah*, one of the holiest moments of Jewish prayer, contains a question that lends itself to contemplation: "Where is the place of God's presence?" But the site of God can't be pinned down, as the continuation of the prayer implies: "Blessed be the presence of *YHVH* from its place." The Hasidic answer to the question is also conducive to contemplation: "Wherever I let God in." Or, the question can be left unanswered, as a Zen koan, a *kavvanah*, a focus of meditation. In the words of one kabbalist, "Concerning everything that cannot be grasped, its question is its answer."

Meditative prayer links us with the cosmos. The *siddur* proclaims that God "in His goodness renews the act of Creation every day, constantly." Chanting these words, I imagine the cycles of life on Earth. Expanding my awareness, I envision the creation of new solar systems in the Milky Way. Somewhere out there right now a star is being born. A clump of matter has attracted gas and dust, grown larger, drawn matter to itself more efficiently until finally the temperature and pressure within are high enough that hydrogen atoms are jammed together and thermonuclear reactions begin. The star turns on and the surrounding darkness is dispelled. Matter turns into light.

Once or twice a year, somewhere in our galaxy, out of a pitch-black cloud of gas and dust a new solar system forms.

And the observable universe contains at least 100 billion galaxies; so 100 solar systems may be forming every second. In that multitude of worlds, many are barren and desolate. Others may be lush, fertile, and animated with forms of life unknown to us but exquisitely adapted to their environment.

With creation unfolding moment by moment, there is no rush: We don't have to read every word on every page. Better to move *slowly* through prayer. With no one counting how many words we have said or not said, our first step is to slow down. The next step is to fully and completely stop—lingering in silence, basking in it. This is difficult for most of us because silence makes us uneasy. Usually our immediate response to silence is to fill it up with anything we can. Even when we are alone, we tend to avoid silence: We take out our phones. If we are with someone else and a moment of silence arises, we immediately squash it with whatever comes to mind, no matter how trivial or inane it may be. We prefer that to the awkwardness of silence. Even lovers rarely share silence for more than a few moments.

But silence can be more profound than words. No words of prayer can adequately praise God, and even the Psalmist, that master of words, said, "To You silence is praise," *Lekha dumiyyah tehillah*. Silence and God share infinity and a pregnant nothingness.

Most of the time my mind is on the move, responding as quickly as it can to all the busyness surrounding me. To begin to fathom the depths of the divine requires a different approach, a ceasing. "Only toward God is my soul silent," *Akh el elohim dumiyyah nafshi*. Entering this silence, I realize it is not mine. It is God's. It envelops and soothes me, loosening the ego, dissolving the story I keep telling myself. The sheer, subtle silence has a texture all its own.

One way to ease into silence is to put aside the book of prayers for a few moments and simply sit and breathe. Become aware of your breathing. Each breath partakes of the breath of creation, the primordial divine breath through which "cosmic space expanded."

Just as God breathed the world into being, so we rejuvenate ourselves through breathing with *kavvanah,* with awareness. There is no need to think profound thoughts. Just to breathe, to inhale an elixir of oxygen and exhale carbon dioxide. Imagine yourself flowing in the stream of life, in the rhythm of the cosmic breath.

The biblical word for "breath" is *neshamah,* as in *nishmat hayyim,* "the breath of life" that God breathed into Adam. But *neshamah* also means "soul." The awareness of breath opens up the dimension of the soul. The soul is not confined within us; rather, we dwell within the soul. Surrounding and permeating us, it is our interface with oneness.

By becoming aware of *neshamah*—breath and soul, soul-breath—we see ourselves more clearly. The *neshamah* exposes what we have hidden away: our secret thoughts and favorite regrets, the pain that we suffer and cling to, the panoply of psychic strategies for propping up the ego. "The *neshamah* of a human is the lamp of God, searching all the innermost chambers." In this verse from Proverbs, what does *neshamah* mean? "Soul"—the powerful beam of awareness, a searchlight into the depths of self. But "breath" also fits—breath as a powerful diagnostic tool. Gentle, deep, relaxed breathing helps you sense where you feel tight and constricted so you can begin to untie the knots—mental, emotional, and physical. As the knots loosen, energy flows more freely and efficiently through the nerves and cells of the body, through the synapses of the mind.

In meditation, thoughts and mental images do not stop, but you let them pass without fixating on them, without becoming attached. Between thoughts, there arises a moment of emptiness, the creative emptiness of *Ayin*. Usually, we rush in to fill this emptiness with words and thoughts; in meditation we savor the gap, "contemplating a little without content, contemplating sheer spirit." A moment of pure awareness, of just being—without thought or action. "Though you do not grasp it, do not despair. The source is still emanating, spreading."

Meditation requires practice and patience. Its fruit, an expansive peace of mind, ripens unexpectedly. Kabbalists call meditation, which is often practiced alone to minimize distractions, *hitbodedut,* "seclusion, aloneness": the temporary separation of the self from the frenetic bustle of daily life.

But meditation is also appropriate for communal prayer in the synagogue. Rabbis should be trained in meditation, so that they can introduce their congregants to spiritual silence. If the rabbi is not a meditator, then a qualified congregant can take the lead.

Silent meditation is appropriate at various points in the traditional service. At the beginning of the service, it can calm us down and usher us into prayer. After a song or a *niggun* (a wordless melody sung over and over), silence lets the music reverberate within us. Various powerful phrases and prayers invite contemplation. As the mind wanders, it is helpful to come back to such phrases, refocus, and begin again.

One of the central prayers of each morning and evening service is the *Shema*. Its opening line proclaims God's oneness: "Hear, O Israel: *YHVH* our God, *YHVH* is One." For centuries, this line has served as a Jewish "mantra," or what is called in Hebrew a *yihud*, a "unification." What does it mean to unify

God? Kabbalists conceive of a contemplative unification of the *sefirot,* the various divine qualities. Hasidim are more simple and direct: "When reciting the word *one* in the unification of the *Shema,* you should intend that there is nothing in the world but the blessed Holy One, whose 'presence fills the whole earth.' Your essence is the soul within, part of God above. Thus only God is. This is the meaning of *one.*"

From a mystical perspective, the command of the *Shema* means: "Hear, O Israel: God is oneness." After chanting these words, a few moments of the service should be devoted to meditating on the word "one," on oneness.

In the Torah scroll, the first line of the *Shema* is written with two of the letters enlarged: *ayin,* which is the final letter of the word *Shema,* and *dalet,* which is the final letter of the word *ehad.* Among the reasons that have been proposed for this oddity is that these two letters spell *ed,* "witness."

Meditation helps us to witness oneness, to pause and appreciate the interwoven wonders of existence. If we dedicate part of the worship service to meditation—to listening to the *alef*—the synagogue will become a spiritual oasis and those praying will discover their own openings to silence, whether these are words from the prayer book or from their own hearts. Some in the congregation may feel somewhat uncomfortable and fill in the silence with whispers or banter or by flipping through pages of the *siddur,* but most will gradually shed their nervousness and try out this new, yet ancient, form of prayer.

The Mishnah describes how "the *hasidim* of old used to wait an hour before praying so that they could focus their mind on God." As the synagogue welcomes meditation back, the words between the silences will breathe more freely and resound, spread new wings of meaning and soar. Prayer will become less rote and less boring. It will recharge us, inspiring us to live out the harmony we have tasted in prayer, to play our part in harmonizing the world.

Just as silence and words enrich one another and need each other, so do stillness and movement. To spend an entire service just sitting down and standing up borders on the sacrilegious. As the Psalmist sings, "All my bones will declare, '*YHVH,* who is like You?'" As the Talmud teaches, one should bend and bow during prayer "until all the vertebrae of the spine loosen."

UNCOVERING THE GEM

There is a strong tendency in religion to accumulate, adding more rules and customs until, finally, they ossify into something venerable called "tradition." This is illustrated by an old Yiddish story about a man who owned a precious gem that he kept hidden in his closet. Toward the end of his life, he gave the gem to his daughter and told her to guard it carefully and pass it on to her oldest child. She obeyed his command and, when she was very old, she handed it on to her son. The son decided to bury it in the backyard and marked the spot with a stone. When his hair turned white, he showed his son the spot. When the son grew old, he forgot what was buried there: He had never seen the gem. But he told his daughter to mark the spot with a stone because it was very important. She did as she was told. Stones were also added by *her* daughter and her grandson and her great-grandson, and on and on for countless generations. By now, an enormous pile of stones had accumulated—and no one knew there was a gem buried underneath.

This is what often happens with ritual: We do what we are "supposed" to do, but we miss the treasure. Even worse, sometimes our preoccupation with the ritual conceals the treasure. So we need to simplify traditional practice. For example, the Torah says that a person should work all week and rest on the seventh day. This is one way we imitate the God of Genesis, who created the world in six days and rested on the seventh. It also reminds us of our slavery in Egypt and our liberation. On Shabbat, we are free not to work. But what constitutes "work"? The Ten Commandments state simply: "You shall do no work." *Lo ta'aseh khol melakhah*. What comprises *melakhah* is not spelled out in the Written Torah, so the rabbis, feeling a need to codify what was permitted and what was forbidden, defined thirty-nine categories of prohibited labor, each of which branches off into numerous subcategories, each of which, in turn, branches off into numerous sub-subcategories. Over the centuries, this has become so complicated that sometimes we forget the essence of Shabbat, which is intended as a day of relaxation and contemplation, as an antidote to materialism. On this day, says the Talmud, we are given an extra soul, *neshamah yeṭerah,* with which we can savor the spiritual dimension of life, nature, family and friends, intimacy, peace, and quiet. Imagine a day when you don't think about money: making it, counting it, spending it. Shabbat is a day to pretend that the mall is closed. The very concept of "the weekend" is, of course, derived from Shabbat, but most American weekends are consumed by shopping and errands. We can restore half of each weekend to its original intent by simply disengaging from materialism.

Praying, singing, studying Torah, and eating delicious food are some of the traditional ways to enhance Shabbat, which begins with lighting the candles and chanting the *qiddush* over wine and the *motsi* over *hallah*. How can we decide which elements of the tradition to follow? It's hard to say. A contemporary *halakhah* should be flexible, drawing on tradition while not confined by it. One rule of thumb is: If there isn't a good reason to reject an element of tradition, give it a chance. Even restrictions that seem extreme may be liberating: not spending time online, not answering the phone, not handling money, not driving a car. Shabbat is an opportunity to recover rare, simple pleasures: walking, reading, being together with those we love. A generous measure of unscheduled time is an essential ingredient of Shabbat. There should be more "yes" than "no" on Shabbat. Whatever our level of observance, there's no need to get obsessed with rules and schedules. Don't bury the gem. Shabbat is a day to appreciate being alive. It is a day to *be*.

TASTING GOD

Eating is a pleasure on any day of the week. By observing the laws of *kashrut*, one ideally brings consciousness to the act of eating. But there are other, simpler ways to be conscious of eating. One is vegetarianism, which, according to Genesis, was the original diet in the Garden of Eden. The first human couple are told: "I give you every seed-bearing plant on the face of the earth and every tree that has seed-bearing fruit; they shall be yours for food." For many Jews, vegetarianism has become "a *kashrut* for our age," not out of asceticism, but rather as an affirmation of life since it embraces the presence of God in all of creation. For some, vegetarianism is too demanding and restrictive, but any spiritual approach should include consuming wholesome food in moderation, and pausing before, during, or after eating to acknowledge from where the food derives and to be aware how we—and the food—fit into the cycle of life. For this, the traditional blessings over food and drink, *berakhot*, are a natural resource.

How can eating, which is such a material act, become spiritual? Certain mystics, feeling that food interferes with spirituality, engage in frequent fasting. A more positive approach recognizes that food provides us with the energy to act, that it can lead to holiness, even if it is not holy in and of itself. But even this approach does not go far enough, according to the Hasidic view: "In everything that you do for the sake of heaven, make sure that you serve God immediately within the act itself. For example, in eating, do not just say that your intention in eating for the sake of heaven is that you should have energy afterward to serve God—although that is also a good intention. Nevertheless, the essence of perfection will be when the deed is immediately holy, raising the sparks. This applies to all human action."

Divinity pervades the universe; its sparks animate every single thing. A physicist would say that energy is latent in subatomic particles and that matter is nothing but bundled energy. From a spiritual perspective, the goal is to raise the sparks, to restore the world to God: to become aware of the interplay between energy and matter; to become aware that every single thing we do or see or touch or imagine is part of the oneness, a pattern of concealed energy.

The material nature of the food that we eat is simply a temporary form assumed by the energy. As we chew food, we begin transforming it back into energy; the taste buds on our tongues mediate a sensation. We taste the spark. "When you eat and drink, you experience enjoyment and pleasure from the food and drink. Arouse yourself every moment to ask in wonder,

'What is this enjoyment and pleasure? What is it that I am tasting?' Answer yourself, 'This is nothing but the holy sparks from the sublime holy worlds that are within the food and drink.'"

This approach toward eating throws a new light on the verse from Psalms: "Taste and see that God is good." Taste the food with *kavvanah,* with spiritual awareness, and you will discover a spark of God. The metaphor reminds us that food is a manifestation of energy, a gift of energy wrapped in matter.

The metaphor, of course, is open to abuse. One Hasidic *rebbe* consumed colossal meals, claiming that he was trying to raise as many sparks as possible! Still, the image of raising the sparks is an effective spiritual technique, not only for eating but for everything else. It transforms religion into a spiritual adventure. In the words of one Hasidic rabbi, "There is no path greater than raising the sparks. For wherever you go and whatever you do—even mundane activities—you serve God."

FROM THE ORDINARY TO THE SPECTACULAR

Energy is everywhere, waiting to be discovered and appreciated. A tangerine in the supermarket is a thing of beauty—*if* I notice it. If I greet the cashier with a friendly smile, our meeting turns from an I-It exchange into an I-Thou encounter. Upon leaving the supermarket, having spent a hundred dollars on food, how can I avoid that homeless man on the sidewalk? Giving him one dollar might raise his spark—and his spirit. The simplest, most mundane activity becomes an opportunity to expand awareness: Rinsing the dishes, I feel the texture of water, essential to life. Making lunches for the kids at 7:30 in the morning, I realize that

I am nourishing them. Driving a carpool of children to school isn't just a chore; now and then, I catch a glimpse of how a child thinks and imagines, what makes him tick.

God is not some separate being up there. She is right here, in the leaf of a tree, in a friend's voice, in a stranger's eye. Isaiah heard the angels singing, "The fullness of the earth is God's presence." In the words of the *Zohar*, "There is no place empty of It." The world is teeming with God. The mystics read the verse "You enliven everything" (*Ve-attah mehayyeh et kullam*), with a twist: "You constitute everything" (*Ve-attah mehavveh et kullam*). Since God is *in* everything, one can serve God *through* everything, by raising the sparks.

Hasidism emphasizes serving God through the physical and the material (*avodah be-gashmiyyut*). This is an earthy, ordinary holiness. The world is God's body, Her primary revelation to us earthbound mortals. Our *halakhah*—the way we walk, the way we live—should reflect this. Ecology is more than a social cause; it is a spiritual principle, a way to honor and preserve the sacred variety of life. According to Genesis, human beings are to "have dominion over the fish of the sea, the birds of the sky, and every living thing that moves on earth." Our mastery of nature has brought us great wealth, comfort, and efficiency. Our technological wizardry has transformed the planet into a global village. Challenging and outwitting gravity, we have left our footprints on the moon. But dominion over nature is not without its curse. By believing that only humans are created in God's image, that we are superior to all other creatures, we have presumed that nature is there for us to do with as we please. We have caused pollution, global warming, ozone-layer depletion, nuclear contamination. We are extinguishing the rainforests—the most luxuriant life system on Earth—at the rate of an acre per second. In the last few decades, we have made extinct approximately one million species. There is still time to avoid environmental catastrophe, but doing so requires redirecting science and technology and reconsidering our self-image as a species. By attuning ourselves to the divine pulse animating all life, we can overcome our estrangement from nature. By exploring and contemplating the origin of the universe, we discover that our evolution is a step in a cosmic dance.

Kabbalah describes the exile of the feminine divine presence (*galut ha-Shekhinah*). This concept originates with Rabbi Akiva, for whom the exile symbolizes God's participation in Israel's suffering. "Wherever Israel has been exiled, *Shekhinah* has been with them." The kabbalists expand this image to convey the cosmic dysfunction that results from the wrenching separation of the masculine and feminine halves of God. The divine divorce interrupts the flow of emanation, vital to the world. Adapting this protean symbol, we could say that patriarchal culture has exiled *Shekhinah* by denying the divinity of nature. As products of that culture, we have lost touch with the Goddess, with the holiness of fertility. When we rediscover the divine presence in the natural world, we redeem *Shekhinah* from Her exile, and She redeems us from alienation.

A contemporary *halakhah* can draw on tradition, yet grow and change day by day. It does not spell out exactly what to do or how to proceed. To a certain degree, the path of this *halakhah* is pathless. Often, we will not follow a precise route, but create a new one. Listening to conscience, heart, and mind, to our needs and the needs of the planet, we integrate these into tradition. We are still "commanded," but the command comes from within.

If *halakhah* becomes too regulated, then the way we move is stiff. An evolving *halakhah* enables us to move freely—or to be perfectly still. Engaging the world spiritually, we realize there is no sharp line between the here-and-now and the ultimate. Looking for the spark, we find that what is ordinary is spectacular. The holy deed is doing what needs to be done *now*.

9. Loving God with the Evil Impulse

If divinity sparkles in everything, what about evil? Is God there, too?

First of all, we should distinguish between unfortunate events and evil. Unfortunate events cause suffering, but no one is directly responsible for them. Examples of such events are natural disasters and many diseases. As Marcus Aurelius wrote, "In the ways of nature there is no evil to be found." Evil arises when someone is responsible for such acts as violent crime, military aggression, or genocide.

God does not "cause" earthquakes, in the sense of "someone up there" deciding that the moment has arrived for two tectonic plates to scrape against each other on a particular stretch of the

San Andreas fault. Natural tragedies are unfortunate, not evil; God is not responsible. Chaos and unpredictability are part of the oneness. But God does manifest in the awesome power of the earthquake, as indicated by the *berakhah* that is traditionally recited on experiencing one: "Blessed are You, *YHVH,* our God, sovereign of the universe, whose power and strength pervade the universe." The same blessing also applies to lightning, thunder, stormy winds, and shooting stars. All these phenomena were traditionally interpreted as *intentional* acts of God, but from a contemporary perspective no separate entity determines what happens when, where, how, or to whom. Since God does not "decide" on a particular course of action, God is not morally responsible for unfortunate events.

What about evil? Here, too, God does not decide or determine what happens: People do. Evil is not a metaphysical reality, a product of the machinations of Satan. Evil is perpetrated by human beings. By blaming it on God or the devil, we absolve ourselves of responsibility.

The Holocaust showed the depth of humanity's capacity for evil. Of the six million Jews murdered, one and a half million were children. Ninety percent of all the rabbis then alive in the entire world were killed. The pain and suffering inflicted by the Nazis does not fit into any divine plan; consequently, the traditional explanation of suffering as punishment no longer works and should be seen for what it is: a survival mechanism that, for centuries, effectively salvaged meaning from tragedy but is obsolete and false for our contemporary consciousness. As the contemporary theologian Edward Feld has written: "The evil ... has no meaning; it is simply there ... a fact, a reality."

Our task is not to explain how God could have "allowed" the Holocaust to happen. Rather, it is to confront evil in the world and in ourselves. Torah commands us to fight the social evil of injustice and to wrestle with the personal evil of selfishness and greed. The Jewish mystics develop a bold theory and technique of transforming the evil impulse (*yetser ha-ra*) into its opposite.

THE SPICE OF DESIRE

Yatsar means "to form or fashion"; *yetser* means "that which is formed or fashioned" in the mind. In other words, it is imagination, which is neither inherently good or evil, *tov* or *ra,* though it can be utilized in either way. In fact, in some Talmudic and midrashic passages *yetser ha-ra,* "the evil impulse," is seen positively: "'God saw everything that He had made, and behold, it was very good.' This refers to *yetser ha-ra.* But is *yetser ha-ra* very good? Yes. Were it not for *yetser ha-ra,* a man would never build a house, marry a wife, have children, or engage in business."

Here *yetser ha-ra* is roughly the libido, expressing itself as desire, passion, and ambition, all of which can create as well as destroy. In the words of an anonymous Talmudic sage, "Whoever is greater than another has a greater *yetser.*" The challenge, as the Mishnah indicates, is to love God not only with the good *yetser* but with the evil one as well. Only in this way can you love God "with all your heart." What this means is not spelled out. Presumably, simply rejecting the evil impulse is not the right approach; we must also try to transform evil into good. This is dangerous, of course, because in attempting such a transformation we may fail and succumb to the evil. Yet within our individual darkness lies an energy yearning to express itself, a spark of life. The libido can be channeled into creativity. As an eighteenth-century preacher said, "*Yetser ha-ra* is like fertilizer for the soul. As manure fertilizes the field, improving its produce, so the fruit of the *tsaddiq* [the righteous person], that is, *mitzvot,* are improved by *yetser ha-ra*. This accords with the well-known principle: 'with all your heart: with both your impulses.'"

Without passion, life is flat, bland, unproductive, and boring. The spiritual path involves more than merely obeying written commands; genuine spirituality is vibrant. By drawing on the depths of one's being, even *yetser ha-ra* can become an essential ingredient of holiness. This radical recipe for a good life appears in a Hasidic interpretation of a passage from the Talmud: "The blessed Holy One said to Israel, 'My children, I have created *yetser ha-ra,* and I have created Torah as its spice.'" The "spice" is an antidote to the poisonous effects of *yetser ha-ra*. But this utilitarian explanation is ignored in the Hasidic reading of the passage, which focuses on the relation between the spice and the main course: "The metaphor does not fit! Spice is added to meat, and the meat is the main dish, not the spice. Yet here God says that Torah is the spice! And so it is: *Yetser ha-ra* is the main thing. One has to serve God with the ecstasy drawn from *yetser ha-ra*."

The ecstasy of *yetser ha-ra* is wild and dangerous, yet indispensable. Constantly keeping it under control is one way to avoid evil, but the price of doing this is repression. Incorporating the *yetser* into our own *halakhah* means that the path we walk may become slippery. To the Hasidic master Nahman of Bratslav, the world is a narrow bridge and while we are on it "the most important thing is not to fear at all." Trying to transform *yetser ha-ra* increases the peril, but the first helpful guideline is not to lose one's balance.

The main component of *yetser ha-ra* is desire. Desire is not inherently bad; as we have seen, it has a positive function. The Midrash, in fact, mentions a special angel "in charge of desire."

Desire keeps us alive: Hunger lets us know that we need to eat; the sexual drive ensures propagation of the species. The problem arises when the desire for *more*—more material things, money, sex, food, attention, status—becomes an end in itself. By fixating on desiring more, we prevent ourselves from enjoying what we already have and who we are right now. The insatiable desire that takes over becomes *yetser ha-ra*.

Since such desire can never be completely fulfilled, it leads inevitably to suffering and neurosis. But by realizing how enslaved we are to desire, we can identify our neuroses: the psychic strategies we have invented to excel in the game of desire. We encounter negative habit patterns and the dark sides of our personalities. Transformation begins when, instead of denying or rationalizing our neurotic behavior, we face it openly and with ruthless compassion. By stepping back and observing the tenaciousness of *yetser ha-ra*, we can loosen its hold on us. Examine it in the light of consciousness, and it provides raw material for new growth; it turns from tyrant to teacher. "Who is wise?" asks the Mishnah. "One who learns from everyone." To which Hasidism adds, "Even from *yetser ha-ra*."

UNTWISTING YOUR THOUGHTS

The mind is not easy to control. If I try to meditate or pray, my concentration is often interrupted by a stray or "strange" thought. How should I respond? The traditional answer was to reject it: "During prayer, if you have a strange thought, then the prayer is unacceptable. You should not assume that only sinful thoughts are forbidden; even thinking of business dealings or other matters is forbidden."

This approach, even when followed rigorously, does not guarantee success. Nahman of Horodenka, who lived in the Ukraine in the eighteenth century, poignantly described his frustrating experience: "When I was very pious, I went every day to a cold *miqveh* [a ritual bath]—so cold that no one in this generation could bear it. When I returned home, the house was very hot—the walls were almost burning—yet I could not feel the warmth for almost an hour. Even so, I could not rid myself of the strange thoughts until finally I was compelled to follow the wisdom of the Ba'al Shem Tov."

The cold shower technique proved ineffective. Nahman found better results when he adopted the new Hasidic approach: "You should believe that 'the whole world is filled with God's presence.' 'There is no place empty of It.' All human thoughts have within them the reality of God. Every thought is a complete creature. If a strange or evil thought arises in your mind while you are engaged in prayer, it is coming to you to be repaired and elevated. If you do not believe this, you have diminished God's existence."

Mental energy is simply another form of the divine energy that permeates everything. Believing in the reality of God means that nothing can be totally rejected. Instead, each strange, twisted thought should be responded to and mended. Each thought is a living creature seeking something from you. Of course, your attempt may fail; instead of untwisting the thought, it may twist you. Aware of this danger, the Ba'al Shem Tov added some conservative advice: "If the way to repair [the thought] does not immediately come to mind, you are then permitted to reject it." The preferred solution, though, is transformation since "the essential purpose of prayer is to elevate and repair all the evil and strange thoughts that arise in your mind." The imaginings of *yetser ha-ra* are not an interruption; they are the main event, the raw material that needs to be processed and refined.

According to the Ba'al Shem Tov's technique, a strange thought can be repaired by identifying the divine quality whose spark is trapped within it. If you feel sexual desire for a married man or woman, neither condemn the desire nor give in to it. Trace the desire back to its archetype, the *sefirah* of love, *Hesed*: "If you have this desire merely because of the single holy spark that is there, how much greater will be the delight if you attach yourself to its source, for there the delight is limitless and incomparable."

By letting go of the particular manifestation of desire, you desire instead the source itself. Such a strategy accords with Plato, who taught that the common people become attached to individual beautiful objects, while the philosopher ascends to a knowledge of beauty itself. At this ultimate stage, fascination with phenomena all but disappears because beauty is eternally unchanging, while beautiful objects are ephemeral.

Similarly with other strange thoughts. A violent impulse arises from an imprisoned spark; it is a skewed expression of the *sefirah* of power. Pride derives from the *sefirah* of beauty: becoming fixated on one's own beauty. "When you bind such thoughts to God, you return them to their root, binding each thought to the quality from which it has fallen. Then the shells fall away and [the thought] is rendered good. God derives great pleasure from this. This can be compared to a prince who has fallen into captivity. When he is brought back to the king, the king is overjoyed, more than for his other son who was always with him."

This raising of the sparks is an act of mystical sublimation. We may differ with the Ba'al Shem Tov about *how* to rechannel our impulses, but his technique is psychologically and spiritually liberating. It frees the energy trapped in unwelcome thoughts, making it available for contemplation and constructive activity. *Yetser ha-ra* is not defeated but rehabilitated; desire is not squelched but redirected. You begin to disengage from particular fixations and experience freedom and spaciousness. By ceasing to want each and every thing, you enable yourself to appreciate what is right here.

There are various strategies for dealing with *yetser ha-ra*. Sometimes its seductive power must be rejected outright. At other times, its wild energy can be sublimated into virtuous action or creativity, thereby channeling its passion into goodness. In certain cases, one can apply a radical insight of the *Zohar*, based on a midrashic teaching about Satan and Yom Kippur. In the original ritual of Yom Kippur (described in Leviticus 16), one goat is sacrificed as a purification offering to God, while a scapegoat bearing the sins of Israel is sent off into the desert for the demon Azazel. According to a remarkable midrash, the goat of Yom Kippur is intended to preoccupy Satan: "They gave him a bribe on Yom Kippur so that he would not nullify Israel's sacrifice."

The *Zohar* develops this one midrashic line and applies it to numerous other rituals and activities. By providing a portion to the demonic force, one ensures that it will be occupied, assuaged, and deterred from interfering in the realm of holiness. What applies to the demonic personality of Satan applies as well to the evil impulse, since according to the colorful Talmudic figure Resh Lakish, "Satan and the evil impulse ... are one and the same." Sometimes, then, the evil impulse can be "bribed" and pacified—given its due so that it no longer hinders virtue.

None of these strategies is always ideal, and none can constantly succeed. It is too easy to justify bribing *yetser ha-ra* by giving in a little when selfish motives determine the whole transaction. Sometimes sublimating an urge is impossible and instead one is overtaken by it and dragged down. Sometimes repressing or subduing *yetser ha-ra* is the best option; but if this becomes our only response, then such piety may be stultifying.

Learning how to wrestle with *yetser ha-ra*—to deal with our desires, passions, and ambition—is a continual challenge. All we can do is to keep trying: to practice loving and living with all our heart, with all our impulses. By employing various strategies, we can gradually learn what works and what doesn't, thereby cultivating practical wisdom. Through failing, we can discover how to succeed, because this is one of those things that "one cannot understand ... unless one has stumbled over them."

10. Israel's Covenant and Other Wisdoms

We have spoken about the ripening of Torah, the evolution of *halakhah*, and the transformation of evil. But how does all this relate to the unique covenant that God made with the people Israel? "Now, if you listen carefully to My voice and keep My covenant, you shall be My treasure among all the peoples. Indeed, the whole earth is Mine, but you shall be My kingdom of priests and a holy nation." According to this basic biblical theme, if the Jewish people commit themselves to living a life of holiness and setting themselves apart from various excesses of the world, God will protect them.

The terms of this covenant fit the pattern of the ancient Near East. Among the Hittites, the king protected his subjects in return for their loyalty and their obedience to a fixed set of obligations. At Sinai, a throng of escaped slaves received God's protection in exchange for promising to obey His commands. The traditional verb for "making" a covenant is *li-khrot,* "to cut." According to Genesis, God's earlier covenant with Abraham was consummated when the patriarch cut several animals in two, and divine smoke and fire passed between the pieces. This fits the Near Eastern pattern, in which the parties to a covenant would pass between an animal that had been severed, as if to say: "If I do not keep the terms of this covenant, may *I* be cut in two."

But now the covenant itself has been cut in pieces: Millions of the Jews Hitler murdered lived according to the Torah, yet their loyalty to the covenant did not secure divine protection. Fire and smoke consumed them.

Even without a Holocaust, the notion of covenant would be problematic today. Since Immanuel Kant in the eighteenth century, it has been argued that a transcendent God cannot dictate authentic moral action; genuine morality emerges only when an individual autonomously follows the dictates of her own reason. Sinai can secure only heteronomous compliance (i.e., under the domination of outside authority); so the demands of such a covenant are ethically questionable or illegitimate.

A COVENANT OF ONENESS

A covenant for today should be based on oneness, not on cutting or division. This covenant cannot be imposed from above, but will emerge and evolve as we discover our interconnection with all of life, since each of us is a unique part of the whole. While this covenant will remind us that ultimately we are not separate, it will not plunge or dissolve us into some vague vastness.

Rather, it signals the work that needs to be done to overcome the many ways in which we are fragmented. By entering into a covenant of oneness, we commit ourselves to increase harmony in the world and to address the selfishness, injustice, and suffering that obscure the oneness.

This covenant is a relationship, not a fixed contract. We need God to remind us that we are part of God, since we constantly forget. And God needs us to mend the fractured world. In Kabbalah, God's need is called "the need on high" (*tsorekh gavoah*). Without us, God is incomplete: The divine sparks remain hidden, the divine potential is unrealized. By mending the world—socially, economically, politically—we mend God and mend what is torn within us, between us, around us.

The traditional covenant challenged the Jewish people to be "a light to the nations," *or la-goyim*. By raising the sparks and mending the world, Jews can model holiness, although complete mending and *tiqqun* will never be achieved—not by them, not even by all peoples working together. Despite the rosy predictions of various traditions, including the Jewish one, the world will never be perfect. Nor will people; so selfishness, greed, cruelty, and suffering will never completely disappear. But we can aim to improve things, to try new approaches, to be creative. As we engage in *tiqqun*, the image we should bear in mind is something more modest and feasible than eternal perfection, an image closer to repairing a car or mending a dress. The *tiqqun* of the world (*tiqqun olam*) is often understood to mean utopian perfection, but its original sense is simply good social policy. Our post-Holocaust covenant demands that we work for this practical type of *tiqqun*. We should better allocate the world's resources and develop the technology to eventually provide food, clothing, shelter, and basic healthcare for all of humanity. We should protect and preserve our planet. We can prove it is possible to survive and flourish without numbing our sense of the holy or diluting our passion for justice.

The covenant binds Jews to Abraham and Sarah, but also to mythical Eve and Adam, then back through billions of generations to the first bubbling of life in the primeval ocean, and finally back to the cosmic seed of creation. The covenant empowers Jews to reveal oneness.

From their history and culture, Jews know what it means to be unique. They recognize—or they should—that each of the seven billion human beings on earth is one of a kind and has fundamental rights. Each family, community, and people has its own history, its own dreams. Jews recently faced extinction; then, in the next moment, they realized a 1900-year-old dream: a Jewish state. While reclaiming their slice of the Middle East, they flexed muscles—political, military, and psychic—that had atrophied in the Diaspora. But how will they ultimately deal with those that have been displaced? To Jews in Israel, Palestinians represent the "other": Their very presence helps define—and threatens—Israeli identity. With slingshots, knives, and firebombs the Palestinians have confronted and annoyed Israel's world-class army, stubbornly refusing to surrender their claim to the same land, which was theirs just yesterday and was the Jews' long ago—and ever since in Jewish memory. Slowly, Palestinians and Israelis must learn to live together and share the Promised Land.

In America, Jews face a different crisis. They have achieved so much and done so well that Jewish leaders and intellectuals fear they may fade into the American mainstream. The danger of the "other" in America—the majority Christian culture—is that its otherness has become familiar, safe, inviting. How do you defend yourself against something so bland and all-encompassing?

The tribe known as Jews is still bound to all humanity. Anti-Semites claim that Jews wield disproportionate power and influence. In a sense, they do. Jews have succeeded financially, socially, intellectually, but their greatest power is an innate need to challenge society and improve it. Jews have reinterpreted Torah, transformed it, even secularized it, but they cannot let go of it—and it will not let go of them. Jews seem programmed to change the world.

Assimilation and intermarriage are inevitable trade-offs of an open society, but the Jewish tribe is tenacious. It will neither die nor dissolve into the culture of the mall, though a century from now it will certainly look very different. Some Jews in the future will lose their precious identity. On the other hand, intermarriage has increased the Jewish gene pool, which is healthy. There is no way to predict how Jewishness will express itself in the generations to come.

PIECES OF THE PUZZLE

Can Jews be loyal to their heritage yet conscious of oneness? No single religion possesses the entire truth. We each have a piece of the puzzle; or rather, we each *are* a piece of the puzzle. So each religious tradition can discover more about the whole from the wisdom and insights of other faiths. Jews should relish their Jewishness but be willing to taste—and be nourished by—the teachings of other religions, without swallowing extreme dogma or claims of exclusivity. In this way, too, Torah ripens.

To whet the appetite, here is a sampling of what other faiths can offer.

Hinduism sings of Brahman, and how It became the world. "In the beginning, there was Existence alone—One only, without a second. It, the One, thought to Itself: 'Let Me be many, let Me grow forth.' Thus, out of Itself, It projected the universe, and having projected the universe out of Itself, It entered into every being. All that is has its self in It alone. Of all things It is the subtle essence. It is the truth. It is the Self. And you are That."

Buddhism teaches that the origin of suffering and unhappiness lies in craving and desire. By reducing our craving, we reduce our suffering. To accomplish this one follows the eightfold path: right understanding, right purpose, right speech, right conduct, right livelihood, right effort, right mindfulness, right contemplation. Following the path requires a moment-by-moment awareness unfettered by material objects or personal biases. Through meditation, one can witness the transient nature of all existence and detach the mind from its various fixations.

Taoism is enigmatic, yet simple: "The names that can be named are not the eternal name." Essence cannot be grasped, only lived. Nothingness is useful: "Thirty spokes converge on a single hub, but it is in the space where there is nothing that the usefulness of the cart lies. Clay is molded to make a pot, but it is in the space where there is nothing that the usefulness of the clay pot lies. Cut out doors and windows to make a room, but it is in the spaces where there is nothing that the usefulness of the room lies. Therefore, benefit may be derived from something, but it is in nothing that we find usefulness."

Islam teaches that the one divine Book has been revealed in various forms throughout history: as Torah, as the Gospels, as the Koran. "For each period there is a book revealed." Moses, Jesus, and Muhammad each conveyed the truth in a way appropriate to their particular age. This relativism is, of course, outweighed by Islam's insistence that the Koran is the final word, that Muhammad is "the seal of the prophets." In Sufism, the mystical stream of

Islam, such exclusivism is at times transcended, for example in the teaching of Ibn Arabi: "My heart is capable of every form: a cloister of the monk, a temple for idols, a pasture for gazelles, the Ka'ba of the pilgrim, the tablets of the Torah, the book of the Koran. Love is my religion. Wherever its camels turn, love is my creed and faith."

Islam and Christianity particularly challenge Jews. Judaism's daughter religions have spread the message of one God throughout the world. Permeating cultures East and West, they have dwarfed their parent faith. Islam and Judaism clash most sharply in the Holy Land, where the fitful peace process cannot succeed without some mutual respect for what each considers holy. In the Middle East and elsewhere, Muslims and Jews must gradually learn to appreciate the wisdom within each other's tradition.

But Christianity is the faith from which Jews find it most difficult to learn, partly because it developed directly from Judaism, partly because of centuries of Christian anti-Semitism. The problem, for many Jews, lies with the very person whom Christians venerate: Jesus.

CAN JEWS LEARN ANYTHING FROM JESUS?

Why should Jews want to have anything to do with Jesus? In his name, Jews have been persecuted and murdered. Several great events in the history of Christianity have been disastrous for the Jews. The Crusades, which sought to liberate the Holy Land from Muslims, meant destruction for scores of European Jewish communities and death for tens of thousands of Jews. The crusading Christians figured that while they were on the way to fight the infidels abroad, why not warm up by attacking the local infidels, the Jews, in the name of avenging Christ's blood? According to the opening book of the New Testament, Matthew, Jews had declared "with one voice ..: 'His blood be on us, and on our children.'" After the Crusades, there were few bright periods for European Jews until the age of the Enlightenment.

Christianity claimed to have supplanted Judaism. Christians were the New Israel and God's covenant had been transferred to them. Centuries of Christian anti-Semitism laid the groundwork for Hitler's extermination of one-third of the Jewish people, so why should any Jew be interested in the life of Jesus and what he taught? The history of Jewish-Christian relations has tainted his image, rendering him impure. Yet for Jews and Christians to live together amicably, they must reevaluate each other's tradition. It is not enough that the Vatican has absolved the Jews of collective guilt for the death of Jesus. Christians should appreciate Torah, rabbinic Judaism, and the eternal renewal of the Jewish people. And Jews should reclaim Jesus.

A GALILEAN *HASID*

I am not talking about what Jesus became—Jesus Christ—but rather about *Yeshua,* the impassioned teacher who died for his vision of Judaism. Jesus was a Galilean *hasid,* someone fervently in love with God, drunk on the divine, unconventional and extreme in his devotion to God and to fellow human beings.

There were other *hasidim* in first-century Palestine, one of whom was strikingly similar to Jesus: Hanina ben Dosa. Hanina lived in Galilee, about ten miles north of Jesus' home town of Nazareth. Like Jesus, he was praised for his religious devotion and healing talents. Once, "Hanina was praying when a scorpion bit him, but he did not interrupt his prayer. His pupils went and found the scorpion dead at the entrance to its hole. They said, 'Woe to the man bitten by a scorpion, but woe to the scorpion that bites [Hanina] ben Dosa.'" Similarly, Jesus said, "Those who believe may step on snakes ... and nothing will harm them." Hanina's prayers were widely regarded as being immediately accepted by God, so he was frequently asked to pray for the sick and those in trouble. According to the Talmud, Hanina cured the son of Gamaliel from a distance; according to the New Testament, Jesus cured the son of the Roman centurion from a distance. Hanina, like Jesus, was known for his poverty and lack of acquisitiveness. Both had no expertise in legal or ritual teachings, but were famous, rather, as miracle workers whose supernatural power derived from their intimacy with God.

Inevitably, tension arises between the *hasid* and the established religious order. The *hasid* is a nonconformist who demands much of himself and of his followers. His intimacy with God, his confidence in the power of his words, and his unrestrained personal authority conflict with the conservative power structure.

Jesus came from Galilee, which made him suspect to both Jewish and Roman authorities since the *Galil* was a hotbed of revolution: Here, the Zealots began their agitation against Rome. The Roman procurator Pontius Pilate killed several Galilean revolutionaries. To the imperial occupiers, any Galilean was a potential troublemaker.

Galilean Jews also had a reputation for a lack of religious observance. Many of their ancestors had been forcibly converted from paganism to Judaism by John Hyrcanus I in the second century B.C.E. In Jesus' time, many pagans still lived there.

The rabbis were suspicious of Galileans, who spoke imperfect Aramaic with a coarse guttural accent. The Talmud sometimes refers to Galileans by the term *am ha-arets,* meaning an ignorant, illiterate peasant. Jerusalem's intellectual elite felt superior towards these unsophisticated provincials.

Exhibiting the chauvinism that was typical of Galilee, Jesus insisted that he was sent to the Jews alone. The twelve apostles were forbidden to proclaim the gospel to Gentiles or Samaritans: Their mission was solely to Israel. In fact, Jesus' disciples later become suspicious of Paul, who sought to preach to the wider world, since Jesus had focused almost exclusively on Jewish affairs.

JESUS' APPROACH TO TORAH

Paul, who never met Jesus in the flesh, taught that Christ had replaced Torah, yet Jesus himself was basically committed to Torah and the *mitzvot.* According to Matthew, Jesus declared,

> Think not that I have come to abolish the Torah and the prophets; I have come not to abolish them, but to fulfill them. For truly, I say to you, till heaven and earth pass away, not an iota, not a dot, will pass from the Torah until all is accomplished. Whoever relaxes one of the least of these commandments and teaches men so, shall be called least in the kingdom of heaven; but he who does them and teaches them shall be called great in the kingdom of heaven.

These words may not be authentic, but Jesus' teaching derives from the Torah. (Similarly, Paul says in the book of Acts, "I assert nothing beyond what was told by the prophets and by

Moses.") Jesus is one of those who are searching for the essence of Torah. When asked by a scribe, "What is the most important commandment?" he answered with two of the *mitzvot*: "Love *YHVH* your God with all your heart, with all your soul, and with all your might" and "Love your neighbor as yourself." Here the Gospel of Mark preserves something that is missing in the other synoptic Gospels. In Luke's and Matthew's accounts of this discussion, there is tension between the scribe and Jesus. In Mark's version, they have a friendly exchange; Jesus and the scribe agree that these are the key *mitzvot*.

Elsewhere, Jesus formulates the essence by paraphrasing "Love your neighbor": "Whatever you wish that people would do to you, do so to them. For this is the Torah and the prophets." As mentioned previously, in the generation preceding Jesus, Hillel had offered a similar principle, but in the negative: "What is hateful to you, do not do to your fellow." Hillel was typical of the rabbis of his day: more down-to-earth, more practical. Jesus was more demanding, more extreme, more hasidic.

It is very difficult to find clear instances of Jesus actually transgressing the Torah. His disciples, not Jesus himself, are accused of disregarding the ritual washing of the hands. This was not a biblical requirement for the laity. It was a purity law spread by the Pharisees, which only in Jesus' time had become a common Jewish practice. Galileans were often lax about purity laws such as this.

Again, it is not Jesus but his disciples who pluck ears of corn on the Sabbath and pull out the kernels. In two secondary sources, we are told that the disciples did not pluck the ears, but removed the kernels by rubbing the ears with their hands. Galileans regarded this as permissible on the Sabbath, while others ruled that it was permissible only when using one's fingers, not the entire hand. Thus, the behavior of the disciples of Jesus the Galilean may have accorded with a Galilean tradition.

There is no indication that Jesus and his disciples ate non-kosher food. According to Mark, Jesus said, "Not what goes into a man defiles him, but the things that come out of a man [i.e., what he says] defile him." Mark interpreted this to mean that Jesus "declared all foods clean," but it is very improbable that this represents Jesus' view. The first generation of Christians did not know that Jesus had "canceled" the food laws, and there is no evidence that Jesus commanded his disciples to ignore them.

Of course, numerous New Testament passages do portray Jesus as being in conflict with the Pharisees. However, these stories reflect the situation in the generations following Jesus when the Gospels were compiled. By now, the early Church had adopted Paul's notion that the laws of Torah had been superseded by Christ, so the dietary and Sabbath laws were openly rejected. Battle lines were clearly drawn between Christianity and rabbinic Judaism, represented respectively in the Gospels by Jesus and the Pharisees.

This is not to claim that everything Jesus said derived from the Torah. Telling one of his followers, "Let the dead bury their dead," ran counter to one of the Ten Commandments: "Honor your father and your mother"—and to Greco-Roman piety as well. The point seemed to be that following Jesus superseded the requirements of piety and Torah. On a number of occasions, Jesus implied that the Mosaic dispensation is inadequate and not final. There is a new age at hand, an eschatological revolution.

At times, Jesus was *more* demanding than Torah. In prohibiting divorce, for example, he went beyond Deuteronomy, which explicitly allows for divorce. From the *mitzvah* "You shall not murder," he concluded that one should not be angry with others since anger can lead to killing. From the *mitzvah* "You shall not commit adultery," he concluded that just glancing lustfully at a married woman is tantamount to adultery. Like a true *hasid,* Jesus was extreme in his ethical demands.

THE THREAT POSED BY JESUS

But if Jesus basically followed the Torah, then why were the civil and religious authorities so upset by him? He associated with sinners, but this, in itself, did not violate tradition, since God, too, yearns for those who have "missed the mark" to return, to engage in *teshuvah*. Various biblical prophets transmitted this message. Joshua told the idolatrous Israelites, "Turn your hearts to *YHVH*"; Ezekiel demanded, "Make yourselves a new heart, a new spirit"; and Malachi transmitted the message from God, "Return to Me, and I will return to you." Eventually, the Midrash portrays God as pleading, "My children, open for Me an opening of *teshuvah* the size of the eye of a needle, and I will open for you openings through which wagons and coaches can pass."

If Jesus promised sinners that by simply believing in him they could gain entrance into the kingdom of heaven, even without repentance, this would have upset many of the normally pious. But still, that doesn't explain the crucifixion. Jesus was arrested, tried, and put to death because he politically threatened both the Roman authorities and the Jewish aristocracy. He had come to Jerusalem at Passover time, the most popular of the three pilgrimage festivals and the one commemorating the liberation from Egyptian bondage. Passover was charged with political significance, and many pilgrims bitterly resented the current pharaoh: the Roman Emperor Tiberius or his local representative, the procurator Pontius Pilate. Trouble was more likely now than at any other season, and normally the Roman procurator came to Jerusalem from Caesarea with extra troops for the garrison.

In Jerusalem, Jesus attacked the money changers in the precincts of the Temple, overturning seats and tables. He looked forward to a new, perfect Temple, which he probably believed would be provided by God from heaven. This "cleansing" of the Temple challenged the priestly aristocracy's political and religious authority, and Jewish leaders concluded that this Galilean should not be allowed to create further trouble.

Jesus may not have said, "I will destroy the Temple" or "I am King of the Jews." He may never have claimed to be the Messiah or the son of God. But his threatening actions and his talk of an imminent kingdom undermined the status quo with Rome. According to Josephus, Herod had similar concerns about John the Baptist, "a good man, who exhorted the Jews to lead righteous lives": "Herod became alarmed. Eloquence that had so great an effect on humankind might lead to some form of sedition. Herod decided therefore that it would be better to strike first and be rid of him before his work led to an uprising."

The Gospel of John portrays the priests as fearful that Jesus' popularity would lead to Roman intervention and disaster: "If we leave him alone like this, the whole populace will believe in him. Then the Romans will come and sweep away our temple and our nation." According to this account, the high priest Caiaphas convinced his colleagues to sacrifice Jesus for the benefit of the entire people: "It is more to your interest that one man should die for the people, than that the whole nation should be destroyed." John's account should not be accepted as historical truth, but it does reflect political considerations that influenced the tragic outcome. Caiaphas' alleged reasoning resonates ironically with the Christian claim that Jesus died for *all* people. In any case, it is fairly certain that Jesus was interrogated by the high priest and then executed on the orders of Pontius Pilate. The charge was sedition or treason.

THE BETRAYAL OF JESUS

Jesus was a charismatic teacher and healer. He did not seek death in Jerusalem, but he pursued with inflexible devotion a path that led to his death, from which he did not try to escape.

Jesus condemned hypocrisy and injustice among his own people and sought to prepare his followers for the coming redemption, for the kingdom of heaven (*malkhut shamayim*). For Jesus, the kingdom was not a pious theory or a far-off promise. It was an immediate reality that could not be denied or evaded. Jewish mysticism later identified the kingdom with *Shekhinah,* the presence of God. Jesus, too, identified kingdom with presence: The kingdom is here and now. Jesus felt compelled to make his fellow Jews aware of this awesome, humbling fact. To enter the kingdom, Jesus said, you must be like a child. Innocence is a window to the infinite, unavailable to the skeptical mind until it pauses and reflects.

Like later *hasidim,* Jesus felt that it was not enough to follow the Torah: One must *become* Torah, living so intensely that one's everyday actions convey an awareness of God and evoke this awareness in others.

Unintentionally, Jesus the Jew founded a new religion. Along with persecuting the Jewish people, Christianity has also spread the Jewish message of monotheism and biblical ethics throughout the world. Jews and Christians need to look at each other anew. Christians should appreciate not only their Jewish roots, but the vitality of contemporary Judaism and the Jewish people—Jesus' people. Jews *can* accept Jesus, not the Jesus of the Church or Jesus Christ, the Messiah, but the Jewish Jesus, a long-lost cousin who for nearly two millennia has been misunderstood and perhaps lonely. By appreciating Jesus as a Jewish teacher, a Jew affirms that the wisdom of Torah manifests in countless, unforeseen ways.

Jesus was a flower of Judaism, cut down in full bloom. Seen through Jewish eyes, he was not the one-and-only son of God. The myth of the son of God explodes into the truth that every human being—every creature, every thing—is an incarnation of God. Jesus should not be idolized. From a Jewish perspective, to turn him into the only son of God is to betray him.

HEREAFTER

11. The End of Days

If, for Jews, Jesus was not the Messiah, then who is? Is the Messiah a being, a divine hero coming to redeem us from alienation and mortality? This notion is the corollary of a more fundamental notion: Someone out there has it all planned out. Mystics, as well as some prominent physicists, believe that there *is* a cosmic intelligence or consciousness. If so, it's not any more out there than right here, not separate from who we are, or what we think and dream. By evolving through us, it becomes aware of itself.

No single person—past, present, or future—is *the* Messiah. But we can help shape a messianic figure by realizing that each of us is one limb of the organism of humanity. The kabbalist Abraham Abulafia saw the messianic age as a "new reality," a time when "each person regards every single human being as a close friend, as one regards each limb of one's body."

Perhaps, through our work of *tiqqun,* through ethical and spiritual activity, we are fashioning Messiah, bit by bit. This kabbalistic perspective resonates with one of Franz Kafka's paradoxical sayings: "The Messiah will come only when he is no longer necessary; he will come only on the day after his arrival." The world is redeemed by justice, by transformative human action. A supernatural Messiah is unnecessary, mere icing on the cake.

Supernatural messiahs and predictions of a messianic age captivate the imagination because the world is so unfair, history is so fickle. Most messianic scenarios spell the end of history in which everything will finally be set right. When the Messiah arrives, we are told, good will finally triumph; evil will be vanquished. That would be nice, but is it really how things work? We shouldn't fool ourselves. The world will never be perfect; society will never be completely just.

THE ULTIMATE FATE OF THE UNIVERSE

According to the most popular current model of cosmology, the universe will keep expanding and accelerating forever, propelled by the mysterious dark energy. However, since the exact nature of dark energy is unknown, we cannot be certain how it will behave. Perhaps over time the expansion will once again slow down.

Assuming that the acceleration will continue indefinitely, there are several scenarios. In one (known as the Big Freeze), existing stars will run out of fuel and be extinguished, one by one. Meanwhile, the supply of cosmic gas will gradually be exhausted, preventing the formation of any new stars. The universe will slowly and inexorably grow darker. Black holes will dominate the universe, though they themselves will eventually disappear, slowly leaking radiation.

In another scenario (known as the Big Rip), all matter in the universe, from galaxies to atoms, will eventually disintegrate into unbound elementary particles and radiation, ripped apart by dark energy and shooting away in all directions.

What about our own neighborhood—the solar system—and the more immediate future, relatively speaking? Well, our sun is about five billion years old—middle-aged and reliable. But in another five billion years, the hydrogen fuel in the sun's core will have nearly run out, and its continuous nuclear reactions will soon cease. The sun will become a red giant. Thermal pressure from within will fade, and nothing will be left to buoy it up. Its core will sag under its own weight, and heat escaping in all directions will make its atmosphere mushroom, engulfing the inner planets and vaporizing Earth. Gradually, most of this atmosphere will fall away, leaving a hot, dense ball of inert matter, which will slowly turn into a cold, dark sphere.

Life will not necessarily come to an end. By then, human beings, or whatever type of intelligent life evolves from us, will have developed the technology to move to another, safer solar system. One prominent theoretical physicist, Freeman Dyson, has suggested that if the universe expands infinitely, life will adapt and evolve, matching its metabolism to the falling temperature of its surroundings: "The pulse of life will beat more slowly as the temperature falls but will never stop." Even if all matter disappears into energy, Dyson theorizes, consciousness may be transferable to a different medium, perhaps resolving itself into energy.

THIS IS IT

Meanwhile, here we are. We still have quite a while until the sun runs out of gas. There will be no absolute *tiqqun,* no final perfection. No one has arranged it all ahead of time. Chance will play a leading role in how things unfold, as it always has. We should learn to negotiate with chance; we should work on mending our own brokenness, our social fabric, our planet as best we can.

Confronted by the here and now, the actual situation is more compelling than any messianic utopia. The interrelatedness of all being is a sheer fact of life: a mystical insight, but also good common sense. The variety-in-oneness that surrounds and includes us demands a flexible approach. We need to learn all we can from the Torah of the past, yet be open to the revelation that is available in this irreplaceable moment. As we translate ancient wisdom into a new idiom, as we improvise with tradition, Torah ripens.

The metaphor of Father in Heaven is inadequate. If God is not a cosmic parent watching over us, then we have to care all the more about each other, the various limbs of humanity.

What kind of God can believe in? The Hebrew word *emunah*, "belief," originally meant trust and faithfulness, both human and divine. Without trusting another person, we cannot love; without trusting others, we cannot build and sustain community. But how can we trust the cosmos, or this God of oneness?

We can trust that we are part of something greater: a vast web of existence constantly expanding and evolving. When we gaze at the nighttime sky, we can ponder that we are made of elements forged within stars, out of particles born in the big bang. We can sense that we are looking back home. The further we gaze into space, the further we see back into time. If we see a galaxy ten million light years away, we are seeing that galaxy as it was ten million years ago, because it has taken that long for its ancient light to arrive here. Beyond any star or galaxy we will ever identify lies the horizon of spacetime, fourteen billion light years away. But neither God nor the big bang is that far away. The big bang didn't happen somewhere out there, outside of us. Rather, we began *inside* the big bang; we now embody its primordial energy. The big bang has never stopped.

And what about God? We can begin to know God by unlearning what we think about God. One of the kabbalistic names for the Infinite is *Nishayon,* "forgetting." One knows God through unknowing, through shedding inadequate conceptions, just as a sculptor cuts away everything that obscures the clarity of the hidden form.

And yet, we cannot capture the divine. God is not an object or a fixed destination. No set of practices, precisely followed, ensures access. There is no definite way to reach God. But then again, you don't need to reach something that's everywhere. God is not somewhere else, hidden from us; God is *right here*, hidden from us. We are enslaved by our routines. Rushing from event to event, from one chore to another, we rarely let ourselves pause and notice the splendor right in front of us. Our sense of wonder has shriveled, victimized by our pace of life.

How, then, can we find God? A clue is provided by one of the many names of *Shekhinah,* the feminine aspect of God, the divine presence. She is called Ocean, Well, Garden, Apple Orchard." She is also called *zot,* which simply means "this." God is right here, in this very moment, fresh and unexpected, taking you by surprise. God is *this.*

Notes

Preface

ever since the big bang In this book I sometimes employ the term "big bang" to include the instant of "cosmic inflation" (discussed in chapter 1), although when cosmologists refer to the big bang they sometimes mean the period immediately following inflation.

"dark energy"... "dark matter" See Neil deGrasse Tyson and Donald Goldsmith, *Origins*, 64–97; Joseph Silk, *On the Shores of the Unknown*, 113–57; Joel R. Primack and Nancy Ellen Abrams, *The View from the Center of the Universe*, 100–11; Richard Panek, *The 4 Percent Universe.*

"the best approximation ..." John D. Barrow and Joseph Silk, *The Left Hand of Creation*, 21.

There were no sound waves ... John Gribbin, *In Search of the Big Bang*, 139–40.

Hoyle gave a radio talk ... The transcript of his talk appears in Fred Hoyle, "Continuous Creation"; the quotations appear on page 568. The following year (1950), Hoyle gave a series of talks for the BBC, and in the concluding one he said, "Broadly speaking, the older ideas fall into two groups. One was that the Universe started its life a finite time ago in a single huge explosion and that the present expansion is a relic of the violence of this explosion. This big bang idea seemed to me to be unsatisfactory even before detailed examination showed that it leads to serious difficulties. For when we look at our own Galaxy there is not the smallest sign that such an explosion ever occurred." See Hoyle, "Man's Place in the Expanding Universe," 420–21.

As for his intention in coining the name, Hoyle later explained in an interview: "The BBC was all radio in those days, and on radio, you have no visual aids, so it's essential to arrest the attention of the listener and to hold his comprehension by choosing striking words. There was no way in which I coined the phrase to be derogatory; I coined it to be striking, so that people would know the difference between the steady state model and the big bang model." See Ken Croswell, *The Alchemy of the Heavens*, 113–14.

Two decades earlier, in 1928, the astronomer and physicist Arthur Eddington had written: "As a scientist I simply do not believe that the present order of things started off with a bang." He later wrote, "Philosophically the notion of an abrupt beginning of the present order of Nature is repugnant to me." See Arthur S. Eddington, *The Nature of the Physical World*, 85; idem, *New Pathways in Science*, 59.

See also Hoyle, *The Nature of the Universe*, 119; idem, *Home Is Where the Wind Blows*, 253–55; idem, as quoted by Alan Lightman and Roberta Brawer, *Origins*, 60; Simon Singh, *Big Bang*, 351–53; Simon Mitton, *Fred Hoyle*, 127–29.

he held a different theory ... According to Hoyle's steady state theory, the universe has been expanding forever without changing its overall appearance, as matter emerges in a process of continuous creation. See Hoyle, *Home Is Where the Wind Blows*, 399–423.

The origin of the cosmos has such grandeur ... Richard Elliott Friedman, *The Disappearance of God*, 264.

"Spacetime tells matter..." See John Archibald Wheeler, *Geons, Black Holes, and Quantum Foam*, 235. Cf. idem, *A Journey into Gravity and Spacetime*, 12: "If spacetime grips matter, telling it how to move, then it is not surprising to discover that matter grips spacetime, telling it how to curve." See also Charles W. Misner, Kip S. Thorne, and John Archibald Wheeler, *Gravitation*, 5.

the infinite God has withdrawn Itself I utilize various pronouns in referring to God in this book. When quoting traditional sources or referring to the traditional conception of God, I often use "He." When referring or alluding to the kabbalistic terms *Shekhinah* (the feminine divine presence) or *Binah* (the Divine Mother), I use "She." When referring to *Ein Sof*, the infinite reality of God beyond categories of gender, I use "It."

the Local Group ... Timothy Ferris, *Coming of Age in the Milky Way*, 175.

Chapter 1

The primordial vacuum ... teeming with virtual particles ... Ferris, *Coming of Age in the Milky Way*, 351–61.

an undifferentiated soup of matter and radiation Steven Weinberg, *The First Three Minutes*, 102.

Kelvin The Kelvin scale begins at absolute zero, the temperature at which molecular energy is at a minimum. This corresponds to a temperature of –273.15° Celsius (or Centigrade). A Kelvin degree is the same size as a Celsius degree.

the universe turned transparent ... "Let there be light!" See Ferris, *Coming of Age in the Milky Way*, 343; Weinberg, *The First Three Minutes*, 7–8; Gerald L. Schroeder, *Genesis and the Big Bang*, 88–90. For a critique of Schroeder's attempt to fit scientific cosmology into the biblical framework of creation, see Friedman, *The Disappearance of God*, 230–34.

"Well, boys, we've been scooped!" Michael D. Lemonick, *Echo of the Big Bang*, 43.

ripples in the fabric of spacetime ... See George Smoot and Keay Davidson, *Wrinkles in Time*, 285; John Gribbin, *In the Beginning*, 37; Joel R. Primack and Nancy E. Abrams, "'In a Beginning ...': Quantum Cosmology and Kabbalah," 68; Friedman, *The Disappearance of God*, 223–27.

We ... are literally made of stardust Technically, stardust makes up about 90 percent of a human body's weight (consisting mostly of oxygen and carbon, along with small amounts of nitrogen, calcium, and phosphorous, and tiny amounts of numerous other elements). About 10 percent of our weight is constituted by hydrogen, all of which originated in the big bang and its aftermath, long before any stars were formed. See Primack and Abrams, *The View from the Center of the Universe*, 89; 324, n. 1.

chunks of material ... collided to form planets Ferris, *Coming of Age in the Milky Way*, 167.

"the best approximation ..." Barrow and Silk, *The Left Hand of Creation*, 21.

beyond the limits of the current theory See Willem B. Drees, *Beyond the Big Bang*; Hubert Reeves, "Birth of the Myth of the Birth of the Universe."

Alan Guth... developed the idea of cosmic inflation... See Alan H. Guth, *The Inflationary Universe*; Singh, *Big Bang*, 477–79. The concept of inflation was first proposed by Guth to solve the problem of magnetic monopoles. He soon realized that the theory also solves two other cosmological difficulties: the "flatness problem" and the "horizon problem." Later it became clear that inflation also explains the variations in density. See Guth, *The Inflationary Universe*, 147–87, 213–43.

Andre Linde ... See Andrei Linde, "The Self–Reproducing Inflationary Universe," 48–55.

a different number of dimensions of spacetime Andre Linde, "Particle Physics and Inflationary Cosmology," 68. Spacetime is usually conceived of as a four-dimensional fabric combining the three dimensions of space and the dimension of time. Each event in the universe represents one point in spacetime.

God "created a universe ..." Andre Linde, "Inflation and Quantum Cosmology," 607.

"Our cosmic home grows ..." Linde, "The Self-Reproducing Inflationary Universe," 55; idem, "Particle Physics and Inflationary Cosmology," 68.

"In a beginning" Primack and Abrams, "'In a Beginning ...'", 71; idem, *The View from the Center of the Universe*, 7.

The Catholic Church endorsed the big bang ... Stephen W. Hawking, *A Brief History of Time,* 49; Singh, *Big Bang*, 360–62.

the singularity disappears See Hawking, *A Brief History of Time,* 137–46.

the dimension of time becomes harder and harder to define Stephen W. Hawking, "The Edge of Spacetime," 14; see Drees, *Beyond the Big Bang*, 55.

Outside of spacetime ... time has no meaning Stephen W. Hawking, "Quantum Cosmology," 651; idem, *A Brief History of Time*, 8.

time itself began at the moment of the big bang Hawking, "Quantum Cosmology," 650; idem, *A Brief History of Time*, 49; Drees, *Beyond the Big Bang*, 55.

"What did God do ..." Augustine, *Confessions*, 11.

"What place, then ...?" Hawking, *A Brief History of Time*, 146; see Paul Davies, *The Mind of God*, 72–73.

God's grammar, the laws of nature See Don Page, in *Origins*, edited by Lightman and Brawer, 409.

For Einstein ... See Norbert M. Samuelson, *Judaism and the Doctrine of Creation*, 237. Samuelson indicates that Einstein's view accords with classical Jewish philosophy.

"Descended from apes!..." Carl Sagan and Ann Druyan, *Shadows of Forgotten Ancestors*, 276.

Shavat va-yinnafash Exodus 31:17.

"The more we know ..." Steven Weinberg, cited in Heinz Pagels, *Perfect Symmetry*, 363–64. Weinberg makes a similar statement at the end of *The First Three Minutes*, 154; see his discussion of the reactions to this statement in *Dreams of a Final Theory*, 255–56. For a wide range of responses to Weinberg from some two dozen leading cosmologists, see Lightman and Brawer, *Origins*, passim.

"God is dead" See Richard L. Rubenstein, *After Auschwitz*. As Eugene Borowitz has pointed out (*Renewing the Covenant*, 36, 41), the traditional God whose death was described by Rubenstein had already been radically reconceived by many Jews.

Yohanan ben Zakkai ... Babylonian Talmud, *Ketubbot* 66b.

to explain the Holocaust ... Rubenstein, *After Auschwitz*, 11, 16, 160. Justin Martyr, the second-century Church father, explained the destruction of the Second Temple as God's punishment of the Jews for the sin of not accepting Christ. See his *Dialogue with Trypho*, 202.

Chapter 2

a contemporary physicist ... Harald Fritzsch, *The Creation of Matter*, 276.

"When you have listened ..." Heraclitus of Ephesus, Fragment 50, in Kathleen Freeman, *Ancilla to the Pre-Socratic Philosophers*, 28. *Logos* means "word" and denotes the rational principle that develops and governs the cosmos.

quarks Quarks were thought up by the Caltech physicist Murray Gell-Mann, who named them with a neologism invented by James Joyce in *Finnegans Wake:* "Three quarks for Muster Mark."

In Genesis ... See Genesis 2:20; *Bereshit Rabbah* 17:4.

the God beyond God This paradoxical phrase appears in the writings of Meister Eckhart. See Bernard McGinn, "The God beyond God."

neither the Bible nor the Talmud ... *Ma'arekhet ha-Elohut*, 82b. The author is unknown.

"The essence of divinity ..." Moses Cordovero, *Shi'ur Qomah,* Modena Manuscript, 206b; see Daniel C. Matt, *The Essential Kabbalah*, 24.

God "exists but not through existence" Moses Maimonides, *The Guide of the Perplexed*, 1:57.

"Know that the description ..." Maimonides, *The Guide of the Perplexed,* 1:58–59.

Ayin ... nihil ... nihts ... nada The Taoist *wu* and the Buddhist *sunyata* and *mu* are similar to Western mystical nothingness but not identical. See Daniel C. Matt, "Varieties of Mystical Nothingness."

David ben Abraham ha-Lavan ... Meister Eckhart ... David ben Yehudah's statement is cited by Gershom Scholem, *Kabbalah*, 95. On Eckhart, see Scholem, *Über einige Grundbegriffe des Judentums*, 74. For a history of *Ayin*, see Daniel C. Matt, "*Ayin*: The Concept of Nothingness in Jewish Mysticism." For other treatments of *Ayin,* see idem, *The Essesntial Kabbalah*, 65–72; Rachel Elior, *The Paradoxical Ascent to God*; Lawrence Kushner, *The River of Light*, 111–34; Rubenstein, *After Auschwitz*, 298–306.

the quantum vacuum Gribbin, *In the Beginning*, 246–47; Arthur Zajonc, *Catching the Light*, 327–28.

a singularity is both destructive and creative Gribbin, *In the Beginning*, 165–66.

"The beginning of existence ..." Moses de León, *Sheqel ha-Qodesh*, 21–22; see Matt, *The Essential Kabbalah*, 70.

A spark of darkness flashed ... *Zohar* 1:15a; see *The Zohar: Pritzker Edition,* Vol. 1, Translation and Commentary by Daniel C. Matt, 107–9.

"When a glassblower wants ..." Shabbetai Donnolo, *Sefer Hakhmoni,* on *Sefer Yetsirah* 1:10; paraphrased by Shim'on Lavi, *Ketem Paz,* 1:49d; see Matt, *The Essential Kabbalah,* 92.

Shaddai: the one who said, "Dai!" ("Enough!") According to BT *Hagigah* 12a (in the name of Resh Lakish), the divine name *Shaddai* alludes to God's original command to limit the expansion of the universe: "I am the one *she,* 'who,' said to the world: *Dai,* 'Enough!'"

With the appearance of the light ... Lavi, *Ketem Paz* 1:124c, on *Zohar* 1:47a; see Matt, *The Essential Kabbalah,* 91.

Heisenberg's uncertainty principle ... Technically, the uncertainty principle applies to momentum (which equals mass times velocity) and position. The formulation in the text assumes that the mass of the particle is known.

"When powerful light is concealed ..." Moses Cordovero, *Pardes Rimmonim,* 5:4, 25d; see Matt, *The Essential Kabbalah,* 91.

whoever thinks that God has an image ... Judah ben Barzilai, *Peirush Sefer Yetsirah,* 14; see Arthur Green, *Seek My Face, Speak My Name,* 215.

This "Nichts of the Jews"... Cited by Scholem, *Über einige Grundbegriffe des Judentums,* 84.

the "sound of sheer silence"... 1 Kings 19:12.

Chapter 3

"The works of creation were made ..." Augustine, *Confessions,* 13:33.

the ten sefirot Literally, *sefirot* means "numerical entities." For further information, see Matt, *The Essential Kabbalah,* 4–11, 38–88.

"With Beginning, through Wisdom ..." Zohar 1:15a; see *The Zohar: Pritzker Edition,* Vol. 1, 110–11; Matt, *The Essential Kabbalah,* 53, 175–76.

Din (Judgment) *Din* is also referred to as *Pahad* (Fear).

"the secret of the possible" On this name of *Shekhinah,* see David ben Judah he-Hasid, *The Book of Mirrors: Sefer Mar'ot ha-Zove'ot,* introduction, 29.

The transition from Goddess to God ... See Tikva Frymer-Kensky, *In the Wake of the Goddesses.*

the cult of Asherah See Raphael Patai, *The Hebrew Goddess*, 34–53; William Dever, *Did God Have a Wife?*, 196–208.

"the revenge of myth" See Gershom Scholem, *On the Mystical Shape of the Godhead*, 140–96; idem, *Major Trends in Jewish Mysticism*, 35..

"nothing at all of Her own" See *Zohar* 1:181a; *The Zohar: Pritzker Edition*, Vol. 3, pp. 86–87, n. 9; pp. 99–100, n. 100.

turned God into ten! See Daniel Chanan Matt, *Zohar: The Book of Enlightenment*, 20.

"like a flame ..." *Sefer Yetsirah* 1:7; *Zohar* 3:70a, 11b.

Desire the wellbeing of your fellow creature ... Moses Cordovero, *Tomer Devorah*, passim; see Matt, *The Essential Kabbalah*, 83, 86.

"Love your neighbor as yourself" Leviticus 19:18.

"When you cleave to the sefirot ..." Joseph ben Hayyim; see Moshe Idel, *Kabbalah: New Perspectives*, 350, n. 333.

Thought reveals itself ... Moses de León, *Commentary on the Ten Sefirot;* see Matt, *The Essential Kabbalah*, 114.

"The inner, subtle essences ..." Isaac the Blind, *Commentary on Sefer Yetsirah*; see Matt, *The Essential Kabbalah*, 113, 201.

"Upon contemplating ..." *Zohar* 3:291b; see *The Zohar: Pritzker Edition*, Vol. 9, 802.

"from our perspective"... *Zohar* 2:176a; 3:141b.

Chapter 4

These objects are impermanent ... Jean Piaget, cited in Robert Ornstein, *The Evolution of Consciousness*, 97–98.

Then one day, we spoke the magic word "I" David J. Wolpe, *In Speech and in Silence*, 31. Wolpe adds: "The first meaning of 'I' is 'not you.'" Cf. Martin Buber, *I and Thou,* 80: "Man becomes an I through a You.... The I-consciousness ... for a long time it appears only woven into the relation to a You, discernible as that which reaches for but is not a You."

Adam originally extended ... Babylonian Talmud, *Hagigah* 12a.

a microbe in the primeval oceans ... Sagan and Druyan, *Shadows of Forgotten Ancestors,* 104–5.

Every cell in the human body ... Ibid., 377–79.

What makes us different? Ibid., 368–69, 400–405.

one-trillionth of the information... Ornstein, *The Evolution of Consciousness,* 169.

Consciousness dawned for a very practical reason ... Richard Dawkins, *The Blind Watchmaker,* 162.

humans redesigned their brains through self-manipulation Daniel C. Dennett, *Consciousness Explained,* 190, 293.

What exactly is consciousness?... Dennett, *Consciousness Explained,* 356, 366, 423, 433; John Searle, *The Rediscovery of the Mind.*

The left hemisphere of our brain ... David Darling, *Deep Time,* 177–79.

a representation of a self Dennett, *Consciousness Explained,* 418, 429.

"Who divorced whom?..." *Zohar* 1:53b; see *The Zohar: Pritzker Edition,* Vol. 1, 297–98.

Rabbi Shim'on bar Yohai ... *Sifrei,* Deuteronomy, 346, citing Isaiah 43:12.

good reasons for preserving the myth ... Dennett, *Consciousness Explained,* 424.

"As a deer pants ..." Psalms 42:2.

"Taste and see ..." Psalms 34:9.

Chapter 5

"showing that God is everything" *Zohar Hadash* 16a (*Midrash ha-Ne'lam*).

the law of gravity implies... See above, p. xiv.

cosmic background radiation See above, chapter 1; and Smoot and Davidson, *Wrinkles in Time.*

The two approaches ... are complementary See Friedman, *The Disappearance of God,* 238–53.

As Einstein wondered ... Ferris, *Coming of Age in the Milky Way,* 385; Davies, *The Mind of God,* 20, 232.

"Something like that ..." Cited in Abraham Pais, *"Subtle Is the Lord ...": The Science and the Life of Albert Einstein,* 131. See Zajonc, *Catching the Light,* 253–70.

Whatever one implants ... *Sha'ar ha-Kavvanah,* attributed to Azriel of Gerona; see Matt, *The Essential Kabbalah,* 110, 200.

"Say to Wisdom ..." Azriel of Gerona, *Perush ha-Aggadot,* 20; see Matt, *The Essential Kabbalah,* 112. The first sentence is from Proverbs 7:4.

"The soul will cleave ..." Isaac of Akko, *Otsar Hayyim* (in manuscript); see Matt, *The Essential Kabbalah,* 112, 201. "More than the calf wants to suck, the cow wants to suckle" is one of five things that the imprisoned Rabbi Akiva taught Rabbi Shim'on bar Yohai; see Babylonian Talmud, *Pesahim* 112a.

"Usually the mind ..." Abraham Isaac Kook, *Orot ha-Qodesh,* 1:268; see Matt, *The Essential Kabbalah,* 125.

Playing with the Hebrew letters of aniy ... See Gershom Scholem, *The Messianic Idea in Judaism,* 214. On the kabbalistic roots of this play on words, see idem, *Major Trends in Jewish Mysticism,* 218.

"Think of yourself as Ayin ..." Dov Baer of Mezhirech, *Maggid Devarav le-Ya'aqov,* 186.

The essence of serving God ... Issachar Baer of Zlotshov, *Mevasser Tsedeq,* 9a–b; see Matt, *The Essential Kabbalah,* 72. Cf. John of the Cross, *Ascent of Mount Carmel* 2:7: "When one is brought to nothing [*nada*]—the highest degree of humility—the spiritual union between one's soul and God will be actualized."

"there is no place empty of It" *Tiqqunei Zohar* 57, 91b.

"Turn away totally ... When you attain the level ..." Dov Baer, *Maggid Devarav le-Ya'aqov,* 224; Levi Yitzhak of Berdichev, *Qedushat Levi,* 71d.

"Thought requires the preconscious ..." Dov Baer, *Or ha-Emet,* 15a. Job 28:12, "Where [*me-ayin*] is wisdom to be found?" is here reinterpreted mystically, as in kabbalistic literature; so now the verse means "Wisdom emerges out of nothingness." (The Hebrew word *ayin* can mean both

"where" and "nothingness.") On *qadmut ha-sekhel,* see Gershom Scholem, *Devarim be-Go*, 2:351–60; Siegmund Hurwitz, "Psychological Aspects in Early Hasidic Literature."

"If we perceive the world ... The foundation of the entire Torah ... In everything they do ..." *Boneh Yerushalayim*, 54; Shneur Zalman of Lyady, *Torah Or*, *Noah*, 11a; *Va-Yetse*, 22b; Dov Baer, *Maggid Devarav le-Ya'aqov,* 24. On the Hasidic concept of *bittul ha-yesh* ("the nullification of *yesh*"), see Rachel Elior, *The Paradoxical Ascent to God*, 143–51.

When you gaze at an object ... Dov Baer, *Maggid Devarav le-Ya'aqov,* 124–25.

"When you sow a single seed ..." Dov Baer, *Maggid Devarav le-Ya'aqov,* 209. This image is widespread. Cf. John 12:24: "Unless a grain of wheat falls into the earth and dies, it remains alone; but if it dies, it bears much fruit." Cf. 1 Corinthians 15:36: "What you sow does not come to life unless it dies." Cf. Koran 6:95: "God is the one who splits the grain of corn and the date-stone. He brings forth the living from the dead."

"when you bring anything ... it is neither chick nor egg" Dov Baer, *Maggid Devarav le-Ya'aqov,* 49, 83–84, 134.

"The 'I' is a thief ..." See Pinhas Zelig Gliksman, *Der Kotsker Rebbe*, 32.

Only when "one's existence ..." Dov Baer, *Maggid Devarav le-Ya'aqov,* 39. Cf. Scholem, *The Messianic Idea in Judaism*, 226–27.

olam In Ecclesiastes 3:11, *olam* could be translated either way: *Gam et ha-olam natan be-libbam*, "God has placed the *olam* within the human mind." This verse serves as a focus for meditation.

"When we become aware ..." *Liqqutei Yeqarim*, 18b–c.

"Before the creation of the universe ..." Shabbetai Sheftel Horowitz, *Shefa Tal*, 3:5, 57b; Hayyim Vital, "On the World of Emanation," in *Liqqutim Hadashim me-ha-Ari u-mi-Maharhu*, 17–18; see Matt, *The Essential Kabbalah*, 93–94.

tsimtsum On *tsimtsum*, see Scholem, *Sabbatai Sevi*, 28–31. On the various interpretations of *tsimtsum*, see Matt, *The Essential Kabbalah,* 91–95; Elior, *The Paradoxical Ascent to God*, 79–91; cf. Michael Wyschogrod, *The Body of Faith*, 98; and from a Christian perspective, Jürgen Moltmann, *God in Creation*, 86–89, 152–57. Here I build on the Hasidic conception.

the vacuum preceding the big bang See above, chapters 1–2.

"You should aim ..." Israel Sarug, *Limmudei Atsilut*, 4d; see Matt, *The Essential Kabbalah*, 97.

The act of shevirah ... Dov Baer, *Maggid Devarav le-Ya'aqov*, 124–27.

An object is symmetrical ... Weinberg, *Dreams of a Final Theory*, 136.

a handful of pencils ... See Ferris, *Coming of Age in the Milky Way*, 313.

this perfect symmetry was broken ... Pagels, *Perfect Symmetry*, 18, 246.

Einstein reimagined it ... On spacetime curvature, see Kip S. Thorne, *Black Holes and Time Warps*, 107–9.

James Clerk Maxwell proved ... Maxwell accomplished this by demonstrating that a flowing electric current has a magnetic field, and a moving magnetic field induces an electric current. This insight led to electric motors and the generation of electricity in power stations. The electromagnetic spectrum includes radio waves, microwaves, and light.

The strong nuclear force ... Within the nucleus, the positively charged protons repel each other—like the repulsion you feel between two magnets if you try to press their north poles together. Without the strong force, the repulsion of the protons would blow the nucleus apart.

the weak nuclear force ... More specifically, the weak force enables a neutron to eject an electron and transform itself into a proton. See Gribbin, *In the Beginning*, 177–78.

a single set of equations ... Gribbin, *In the Beginning*, 179.

soon the forces become distinct ... See Barrow and Silk, *The Left Hand of Creation*, 86.

a Platonic perspective See Weinberg, *Dreams of a Final Theory*, 195.

"Physicists, in identifying ..." Ferris, *Coming of Age in the Milky Way*, 334.

"When we become aware ... God does not derive ..." *Liqqutei Yeqarim*, 18b–c; Ze'ev Wolf of Zhitomir, *Or ha-Me'ir*, *Ki Tissa*, 26d–27a.

"God wants to be served ..." *Tsavva'at ha-Ribash*, in *Shivhei ha-Besht*, 215.

"With the concealment of the light ..." See above, chapter 2.

"The universe is built ..." Paul Valéry, cited in Ferris, *Coming of Age in the Milky Way*, 301.

a drop of water ... the ocean and the wave ... See Rubenstein, *After Auschwitz*, 298–99; Idel, *Kabbalah: New Perspectives*, 67–70.

just seeing another person ... See Emmanuel Levinas, *Nine Talmudic Readings*, 47–48: "The epiphany of the other person is ... my responsibility toward him: seeing the other is already an obligation toward him."

The purpose of the marriage ... Abraham Abulafia, *Mafteah ha-Tokhahot;* see Matt, *The Essential Kabbalah*, 21.

Chapter 6

"O people of the Book!..." Koran 5:16.

"That which is plain ... I would like to see ..." Judah Halevi, *Kuzari* 3:35.

"If a case is too baffling ..." Deuteronomy 17:8–11.

613 mitzvot Babylonian Talmud, *Makkot* 23b. The 248 limbs are enumerated in Mishnah, *Ohalot* 1:8. The total number of days plus limbs, 613, actually preceded the precise enumeration of the *mitzvot*. Only after the Talmud was composed did anyone attempt to work out an exact list of 613 commandments. Since many of the *mitzvot* apply to various rituals surrounding the Temple in Jerusalem, only about ten percent of the 613 are now practical. See Maimonides, *Sefer ha-Mitzvot,* at the end of his list of the positive commandments.

"One who walks in righteousness ..." Isaiah 33:15–16.

"It has been told to you ..." Micah 6:8.

"Observe what is right ..." Isaiah 56:1 (actually written by a later prophet).

"Seek Me and live" Amos 5:4. The list of verses in *Makkot* (23b–24a) concludes with Habakkuk 2:4: "The righteous shall live by his faith." Another single, all-encompassing verse (Proverbs 3:6) is cited by Bar Kappara in Babylonian Talmud, *Berakhot* 63a: "What is a brief passage upon which all the essentials of Torah depend? 'In all your ways know God, and He will smooth your paths.'"

Hillel's response ... Babylonian Talmud, Shabbat 31a.

"Whatever you wish ..." Matthew 7:12.

"Love your neighbor ..." Leviticus 19:18.

Elsewhere in the Gospels ... Mark 12:28–34, citing Deuteronomy 6:5. Here Jesus and one of the scribes have a friendly conversation and find themselves in agreement.

"a central principle ... even if God ..." *Sifra, Qedoshim* 4:12, 89b; Babylonian Talmud, *Berakhot* 61b.

"Whatever you do ..." *Sifrei,* Deuteronomy, 41.

"Not a hoof ..." Exodus 10:26. Thanks to Rabbi Jeremy Milgrom for showing me this reading of the verse.

Rabbi David ibn Zimra was asked ... David ben Solomon ibn Abi Zimra, *She'elot u-Teshuvot ha-Radbaz,* 4:3a, responsum no. 1087.

"One who fulfills ..." *Or Torah,* 147a, citing Proverbs 3:6; see above, note on "*Seek Me and live.*"

"When you are walking ..." *Tsavva'at ha-Ribash,* in *Shivhei ha-Besht,* 215.

There was a man ... *Zohar* 2:176a–b; see *The Zohar: Pritzker Edition,* Vol. 5, 531–34; Matt, *The Essential Kabbalah,* 207.

Chapter 7

Torah Qedumah... Primordial Torah ... See Scholem, *On the Kabbalah and Its Symbolism,* 41.

"The tao ..." The opening lines of the *Tao te Ching,* attributed to Lao Tzu, the sixth-century B.C.E. founder of Taoism. *Tao* originally means "road" or "pathway" and comes to mean "the natural or correct way something is done," also "teaching," "moral truth," "the way the whole universe operates."

God gazed into the Torah ... *Bereshit Rabbah* 1:1.

Einstein's view ... See above, chapter 1.

"an unripe fruit of heavenly wisdom" *Bereshit Rabbah* 17:5, in the name of Rabbi Avin: *Novelet hokhmah shel ma'lah torah.*

"Fire is one-sixtieth ..." Babylonian Talmud, *Berkahot* 57b.

"But isn't the Torah ..." Moses de León, *Sefer ha-Rimmon*, 107, 326; see Matt, *The Essential Kabbalah,* 145, 213.

"without vowels ..." Bahya ben Asher, *Commentary on the Torah,* Numbers 11:15; see Idel, *Kabbalah: New Perspectives,* 213–14; Matt, *The Essential Kabbalah,* 146, 213.

"The Torah scroll may not be vowelized ..." Jacob ben Sheshet, *Meshiv Devarim Nekhohim*, 107. Cf. Moshe Idel, "Infinities of Torah in Kabbalah," 146–47. According to Jacques Lacan, the analyst repunctuates the patient's discourse.

according to one kabbalistic theory ... See Scholem, *On the Kabbalah and Its Symbolism*, 77–86; idem, *Major Trends in Jewish Mysticism,* 179. Joachim of Fiore, the twelfth-century Cistercian monk, taught a similar doctrine in which various manifestations of the divine personality (Father, Son, and Holy Spirit) appear as symbols of successive cosmic eons.

their domain of validity Thorne, *Black Holes and Time Warps,* 84.

"Every day that you study Torah ..." *Tanhuma,* ed. Solomon Buber, *Yitro,* 7.

"we have deeds of kindness" *Avot de-Rabbi Natan,* Version A, 4.

pay compensation ... Mishnah, *Bava Qamma* 8:1.

"It is a mitzvah ... Do not think ..." Jacob ben Sheshet, *Sefer ha-Emunah ve-ha-Bittahon,* 364, 370. Cf. Matt, "The Mystic and the *Mizwot,*" 379.

One Talmudic source ... Another ... Babylonian Talmud, *Gittin* 60a.

"I am YHVH your God ..." Exodus 20:2–3.

"from the mouth of Power"... Babylonian Talmud, *Makkot* 24a.

a later, Hasidic view ... Quoted in the name of Menahem Mendel of Rymanow. See Scholem, *On the Kabbalah and Its Symbolism,* 29–31; Kushner, *The River of Light,* 16, 59, 62; Arthur Green, *Radical Judaism,* 90–92.

the white space behind the letters ... Scholem, *On the Kabbalah and Its Symbolism,* 48–50; Idel, "Infinities of Torah," 145. The Talmudic passage is recorded in the name of Shim'on ben Lakish in Jerusalem Talmud, *Sheqalim* 6:1.

"That which the prophets ..." *Shemot Rabbah* 28:6; 47:1.

"One, God has spoken ..." Psalms 62:12. In *Mekhilta de-Rabbi Shim'on bar Yohai* on Exodus 20:1, the verse from Psalms is cited to explain how God could have simultaneously spoken the two different versions of the Ten Commandments in Exodus 20 and Deuteronomy 5. Also cited is Jeremiah 23:29: "'Is not My word like fire?' says *YHVH*." The Midrash concludes: "Just as fire splits into many sparks, so one divine utterance issues in many texts."

the sound of sheer silence 1 Kings 19:12.

"The Torah that one studies ..." *Qohelet Rabbah* 11:8.

"When the blessed Holy One ..." *Shemot Rabbah* 29:9, which reads: "no ox lowed."

"one cannot acquire wisdom ..." *Bemidbar Rabbah* 1:7; cf. Babylonian Talmud, *Eruvin* 54a.

"The divine word spoke ... The blessed Holy One appeared ..." *Pesiqta de-Rav Kahana* 12:24.

Abraham experienced revelation ... See *Bereshit Rabbah* 61:1: Rabbi Shim'on bar Yohai said, "The blessed Holy One appointed Abraham's two kidneys [the seat of the emotions] to be like two rabbis, and they flowed and taught him Torah and wisdom."

"Every single person should say ..." *Seder Eliyyahu Rabbah* 25, p. 127.

"Know that all is God ..." Menahem Nahum of Chernobyl, *Me'or Einayim, Qedoshim,* 47a, citing Psalms 81:10.

"Who is wise? ..." Mishnah, *Avot* 4:1, in the name of Shim'on ben Zoma.

"A Torah will come forth ..." Isaiah 51:4.

"I will place My Torah ..." Jeremiah 31:32.

Chapter 8

what philosopher Emil Fackenheim calls the 614th commandment See Emil L. Fackenheim, *The Jewish Return into History,* 19–24.

"One who writes down halakhot ..." Babylonian Talmud, *Temurah* 14b.

"Love your neighbor ... Do justice ..." Leviticus 19:18; Micah 6:8.

"Do not oppress ..." Exodus 23:9.

"When a stranger resides ..." Leviticus 19:33–34, in the same chapter as "Love your neighbor."

"If a man lies ..." Leviticus 18:22; 20:13.

liturgy ... replaced sacrificial worship ... Babylonian Talmud, *Berakhot* 26b.

Let us bless ... See Marcia Falk, *The Book of Blessings.* All the prayers appear in distinct Hebrew and English versions.

cosmic background radiation ... See Primack and Abrams, "'In a Beginning ...': Quantum Cosmology and Kabbalah," 66; and above, chapter 1.

the science closest to theology Barrow and Silk, *The Left Hand of Creation,* 226.

"humanity's first verifiable creation story" Primack and Abrams, "'In a Beginning ...': Quantum Cosmology and Kabbalah," 66.

the divine language is energy ... See above, chapter 1.

to develop intimacy with what seemed beyond See Green, *Seek My Face, Speak My Name,* 22–23.

"Concerning everything that cannot be grasped ..." Shim'on Lavi, *Ketem Paz,* 1:91a.

a star is being born ... Sagan and Druyan, *Shadows of Forgotten Ancestors,* 12–13.

"To You silence is praise" Psalms 65:2. See Maimonides, *The Guide of the Perplexed,* 1:59.

"Only toward God is my soul silent" Psalms 62:2. Cf. Psalms 37:7: "Be silent toward God."

"cosmic space expanded" See above, chapters 1–2.

"The neshamah of a human ..." Proverbs 20:27.

"contemplating a little ... Though you do not ..." See above, p. 40.

hitbodedut ... See Wolpe, *In Speech and in Silence,* 191; Idel, "*Hitbodedut* as Concentration in Ecstatic Kabbalah."

"When reciting the word one ..." *Liqqutei Yeqarim,* 12b. The biblical citation is from Isaiah 6:3.

ed, "witness" See Jacob ben Asher, *Ba'al ha-Turim,* on Deuteronomy 6:4; *Zohar* 3:236b.

"the hasidim of old ..." Mishnah, *Berakhot* 5:1.

"All my bones ... until all the vertebrae ..." Psalms 35:10; Babylonian Talmud, *Berakhot* 28b.

"You shall do no work" Exodus 20:10; Deuteronomy 5:14.

"I give you every seed-bearing plant ..." Genesis 1:29.

"a kashrut for our age" Green, *Seek My Face, Speak My Name,* 87–89.

"In everything that you do ..." Jacob Joseph of Polonnoye, *Toledot Ya'aqov Yosef, Mishpatim,* 59b.

"When you eat and drink ..." Alexander Susskind, *Yesod ve-Shoresh ha-Avodah,* 7:2, 60a; see Matt, *The Essential Kabbalah,* 150.

"Taste and see ..." Psalms 34:9.

"There is no path greater ..." Levi Yitshak of Berdichev, *Qedushat Levi, Vayeshev,* 26b; see Matt, *The Essential Kabbalah,* 151.

"The fullness of the earth ..." Isaiah 6:3, usually rendered: "The whole earth is filled with His glory [or: presence]," but technically the Hebrew word *melo* is not an adjective meaning "filled," but rather a noun meaning "fullness."

"There is no place empty ..." *Tiqqunei Zohar* 57, 91b.

The mystics read the verse ... Elijah de Vidas, *Reshit Hokhmah, Sha'ar ha-Qedushah,* 7:185a, citing Nehemiah 9:6.

"have dominion ..." Genesis 1:28.

"Wherever Israel has been exiled ..." *Mekhilta, Pisha,* 14; Babylonian Talmud, *Megillah* 29a.

Chapter 9

distinguish between unfortunate events and evil See Lawrence Kushner, *God Was in This Place and I, i Did Not Know*, 61–63.

"In the ways of nature ..." Marcus Aurelius, *Meditations*, 2:17.

"Blessed are You ..." *Mishnah, Berakhot* 9:2; cf. Babylonian Talmud, *Berakhot* 58b–59a. Customarily, the blessing over lightning is "... who enacts the act of Creation."

"The evil ... has no meaning ..." Edward Feld, *The Spirit of Renewal*, 84–85, 139–40.

"'God saw everything ..." *Bereshit Rabbah* 9:7, citing Genesis 1:31. See Daniel Boyarin, *Carnal Israel*, 61–76.

"Whoever is greater than another ..." Babylonian Talmud, *Sukkah* 52a.

"with all your heart" Mishnah, *Berakhot* 9:5.

"Yetser ha-ra is like fertilizer for the soul ..." Eliezer Lipman, *Tal Orot im Migdal David*, 162b.

"The metaphor does not fit ..." Cited in the name of the Maggid of Mezherech by Abraham Hayyim of Zlotshov, *Orah la–Hayyim,* 1:13a. The Talmudic passage appears in Babylonian Talmud, *Qiddushin* 30b.

a special angel ... *Bereshit Rabbah* 85:8.

"Who is wise? ..." Mishnah, *Avot* 4:1; Jacob Joseph of Polonnoye, *Tsofnat Pa'neah*, 8a.

"During prayer ..." Zevi Hirsch Koidonover, *Qav ha-Yashar*, chapter 8, 10a.

"When I was very pious ..." *Shivhei ha-Besht,* ed. Mintz, p.112.

"You should believe ..." Jacob Joseph of Polonnoye, *Ben Porat Yosef,* 43a. The passages cited at the beginning of the quotation are from Isaiah 6:3 and *Tiqqunei Zohar* 57, 91b.

"If the way to repair ..." This piece of advice may originate with Jacob Joseph, who transmits this teaching.

"If you have this desire ..." Jacob Joseph of Polonnoye, *Toledot Ya'aqov Yosef, Eqev,* 172a.

accords with Plato ... See Wyschogrod, *The Body of Faith,* 132.

"When you bind ..." *Keter Shem Tov*, 37c.

"They gave him a bribe on Yom Kippur..." *Pirqei de-Rabbi Eli'ezer*, chapter 46. Cf. *Sifra, Shemini* 1:3, 43c.

By providing a portion to the demonic force... See *The Zohar: Pritzker Edition*, Vol. 5, 102, n. 291.

"Satan and the evil impulse..." Babylonian Talmud, *Bava Batra* 16a.

"one cannot understand..." Babylonian Talmud, *Gittin* 43a, speaking of "words of Torah."

Chapter 10

"Now, if you listen ..." Exodus 19:5–6.

the Hittites The chief power and cultural force in Western Asia from 1400 to 1200 B.C.E. They were also one of the first peoples to smelt iron.

According to Genesis ... Genesis 15.

Since Immanuel Kant ... See Arnold Eisen, "Covenant," 110.

not a fixed contract See Borowitz, *Renewing the Covenant,* 31.

good social policy See, e.g, Mishnah, *Gittin* 4:2–3; *Sifrei*, Deuteronomy, 113.

to reveal oneness See Green, *Seek My Face, Speak My Name,* 121.

Assimilation and intermarriage ... See Neil Gillman, *Sacred Fragments,* xxii.

"In the beginning ..." Chandogya Upanishad, in *The Upanishads*, 46.

"Thirty spokes converge ..." Lao Tzu, *Tao Te Ching*, 59, 70.

"For each period ..." Koran 13:38.

"My heart is capable ..." Ibn al-Arabi, *Tarjuman*; see Annemarie Schimmel, *Mystical Dimensions of Islam*, 272.

According to ... Matthew ... Matthew 27:26.

"Hanina was praying ..." *Tosefta, Berakhot* 3:20. In the following pages I am drawing on Geza Vermes, *Jesus the Jew*; E. P. Sanders, *Jesus and Judaism*; idem, "The Life of Jesus." See also Amy-Jill Levine, *The Misunderstood Jew*.

"Those who believe ..." Mark 16:18.

Exhibiting the chauvinism ... Matthew 10:5; 15:24.

Think not that I have come ... Matthew 5:17–19.

"I assert nothing beyond ..." Acts 26:22.

the Gospel of Mark Mark 12:28–34.

"Whatever you wish ..." Matthew 7:12.

It is very difficult to find ... See Sanders, *Jesus and Judaism*, 245–69; idem, "The Life of Jesus," 70–73.

His disciples, not Jesus himself ... Mark 7:1–8.

Again, it is not Jesus ... Matthew 12:1–8.

In two secondary sources ... See David Flusser, *Jewish Sources in Early Christianity*, 22.

"Not what goes into a man ..." Mark 7:15.

there is no evidence ... See Sanders, "The Life of Jesus," 72.

"Let the dead bury their dead" Matthew 8:22.

"missed the mark" The Hebrew root *ht'*, "to sin," means, in the *hiph'il* conjugation, "to miss the mark."

Joshua ... Ezekiel ... Malachi Joshua 24:23; Ezekiel 18:31; Malachi 3:7.

"My children, open for Me ..." *Shir ha-Shirim Rabbah* 5:3.

the three pilgrimage festivals The other two pilgrimage festivals are *Shavu'ot* (Pentecost, or the Festival of First Fruits) and *Sukkot* (the Festival of Booths).

Trouble was more likely ... Sanders, "The Life of Jesus," 75.

overturning seats and tables Though, as we have seen, Jesus condemned anger, he himself exhibits a fair amount of righteous anger.

This "cleansing" of the Temple ... Mark 11:15–17; John 2:14–17.

According to Josephus ... Josephus, *Antiquities of the Jews* 18:117–18, discussed by Vermes, *Jesus the Jew,* 50–51.

"If we leave him alone ..." John 11:47–50.

Chapter 11

a "new reality ..." Abraham Abulafia, *Mafteah ha-Shemot*, 48a; idem, *Or ha-Sekhel*, 37b. See Abraham Berger, "The Messianic Self-Consciousness of Abraham Abulafia," 58–59.

"The Messiah will come ..." Franz Kafka, *Parables and Paradoxes*, 81.

But in another five billion years... Thorne, *Black Holes and Time Warps,* 159; Darling, *Deep Time,* 139–141.

"The pulse of life ..." Freeman Dyson, *Infinite in All Directions*, 110–21.

We still have quite a while until the sun runs out of gas Actually, the Earth will become uninhabitable long before then, as the sun increases in luminosity and the oceans evaporate.

outside of us... See Edwin Turner, in Lightman and Brawer, *Origins,* 320.

Nishayon, "forgetting" See Matt, *The Essential Kabbalah*, 63, 180–81.

just as a sculptor ... Pseudo-Dionysius Areopagite, *The Mystical Theology,* chapter 2. See Abraham Isaac Kook, "Pangs of Cleansing," in Matt, *The Essential Kabbalah,* 32–35.

Glossary

absolute zero: The temperature at which molecular energy is at a minimum. This corresponds to a temperature of 0° Kelvin or –273.15° Celsius (or Centigrade).

alef: The first letter of the Hebrew alphabet; the beginning of divine and human speech.

antiparticle: A particle with identical mass as another particle, but with equal and opposite electrical charge. When a particle encounters its own antiparticle, both annihilate into radiation.

atom: The basic unit of ordinary matter, made up of a tiny nucleus (consisting of protons and neutrons) surrounded by orbiting electrons.

Ayin: "Nothingness," the creative "no-thingness" of God, out of which all being flows.

Ba'al Shem Tov: "Master of the Good Name," the title of Israel son of Eli'ezer, charismatic founder of Hasidism (ca. 1700–1760).

big bang: The cosmological theory according to which the universe began in a state of extremely high density and temperature about fourteen billion years ago and has been expanding, thinning out, and cooling ever since.

Binah: "Understanding," the third *sefirah*, the Divine Mother who gives birth to the seven lower *sefirot*.

black hole: A region of spacetime with such enormous gravitational force that nothing, not even light, can escape. Massive stars, after exhausting their nuclear fuel, collapse under their own weight to form black holes. All massive galaxies harbor black holes at their center.

cosmic background radiation: The sea of photons produced everywhere in the early universe. This remnant afterglow of the big bang still pervades all space and was discovered by accident in New Jersey in 1964.

cosmology: The study of the evolution and structure of the universe.

dark energy: Energy that is invisible and undetectable by any direct measurement, and that makes space expand. Although it seems to constitute nearly 70 percent of the mass-energy of the universe, its nature is unknown.

dark matter: An enigmatic form of matter that emits no eletromagnetic radiation. Based on the gravitational force that it exerts on visible matter, it apparently comprises the bulk of all matter in the universe.

Ein Sof: "There is no end," that which is boundless, the Infinite; the ultimate reality of God beyond all specific qualities and attributes. The God beyond God.

electron: A subatomic particle with a negative electric charge that orbits the nucleus of an atom.

galaxy: A large group of stars that are bound together by gravitation.

Gevurah: "Power," the fifth *sefirah*; also called *Din*, "Judgment."

halakhah: "Walking, going," the discipline of Jewish religious practice.

Hasidism: A popular religious movement that emerged in the second half of the eighteenth century in Eastern Europe.

Hesed: "Love," the fourth *sefirah*, balancing *Gevurah*.

Hokhmah: "Wisdom," the second *sefirah*, the primordial point of emanation.

Kabbalah: "Receiving," that which is handed down as tradition; the esoteric teachings of Judaism.

kavvanah: "Intention," focused awareness in prayer or meditation.

Kelvin: A scale of temperature that begins at absolute zero, the temperature at which molecular energy is at a minimum. This corresponds to a temperature of –273.15° Celsius (or Centigrade). A Kelvin degree is the same size as a Celsius degree.

Keter: "Crown," the first *sefirah*; also called *Ayin*, "Nothingness."

Mishnah: The collection of oral teachings arranged and revised at the beginning of the third century by Rabbi Judah Ha-Nasi; the earliest codification of Oral Torah.

mitzvah **(pl.** ***mitzvot*****):** "Commandment," one of the 613 precepts of the Torah; by extension, good deed.

neutron: A subatomic particle that together with the proton makes up the atomic nucleus. The neutron has no electrical charge and is composed of three quarks.

nucleus: The central part of an atom, consisting of protons and neutrons.

Oral Torah: The interpretation of the Written Torah.

photon: A subatomic particle that transmits electromagnetic force. Light consists of a stream of photons.

proton: A subatomic particle that together with the neutron makes up the atomic nucleus. The proton has a positive electrical charge and is composed of three quarks.

quantum: The smallest quantity of radiant energy.

quantum fluctuation: Continuous variations that can cause particles to appear and disappear in the vacuum of "empty" space. Some theories hold that the entire universe was created out of nothing in a quantum fluctuation.

quark: One of the fundamental particles, out of which protons and neutrons are made. Quarks were thought up by the Caltech physicist Murray Gell-Mann, who named them with a neologism invented by James Joyce in *Finnegans Wake*: "Three quarks for Muster Mark."

sefirah **(pl.** ***sefirot*****):** One of ten aspects of God's personality, according to the Kabbalah.

Shekhinah: "Presence," the tenth *sefirah*, divine immanence, female partner of *Tif'eret*.

Shema: "Hear," one of the central daily prayers in Judaism, which begins with Deuteronomy 6:4: "Hear, O Israel: *YHVH* our God, *YHVH* is One!"

shevirah: "Breaking" of the vessels, the shattering of primordial unity.

siddur: "Arrangement," the traditional Jewish prayer book.

singularity: A point in spacetime with infinite density. The laws of physics break down at a singularity.

spacetime: A four-dimensional fabric combining the three dimensions of space and the dimension of time. Each event in the universe represents one point in spacetime.

symmetry: "The same measure," harmonious proportions. An object is symmetrical if it looks the same from different points of view.

Talmud: The body of teaching comprising the Mishnah and Gemara (commentary and discussions on the Mishnah by scholars of the third-to-fifth centuries).

Tif'eret: "Beauty," the sixth *sefirah*, male partner of *Shekhinah*.

tiqqun: "Repair," "mending," restoring the fractured unity of existence through ethical and spiritual activity.

tsimtsum: "Contraction" or withdrawal of divinity, by which God made room for the world to exist.

vacuum: A state of minimum energy. Empty space is often referred to as the vacuum, though it still has a minimum energy content.

virtual particle: A particle that cannot be directly detected or captured, but whose existence does have measurable effects. These particles are created in the vacuum in pairs and annihilate quickly.

Written Torah: The first five books of the Bible: Genesis through Deuteronomy.

yetser ha-ra: "The evil impulse."

YHVH: The ineffable name of God, derived from the Hebrew root *hvh*, "to be."

Zohar: "Radiance," the masterpiece of Kabbalah, probably composed in large part by Moses de León in thirteenth-century Spain and attributed by him to the circle of Rabbi Shim'on bar Yohai, who lived in Palestine in the second century.

Bibliography

Abraham Hayyim ben Gedaliah of Zlotshov. *Orah la–Hayyim*. Jerusalem: Y. Wolf and Y. D. Shtitzberg, 1960.

Abrams, Nancy Ellen. *A God That Could Be Real: Spirituality, Science, and the Future of Our Planet*. Boston: Beacon Press, 2015.

Abulafia, Abraham. *Mafteah ha-Shemot*. Jewish Theological Seminary Manuscript 843.

———. *Or ha-Sekhel*. Columbia University Library Manuscript X893 Ab 92.

Adams, Fred. *Our Living Multiverse: A Book of Genesis in 0+7 Chapters*. New York: Pi Press, 2004.

———, and Greg Laughlin. *The Five Ages of the Universe: Inside the Physics of Eternity*. New York: The Free Press, 1999.

Alexander Susskind ben Moses of Grodno. *Yesod ve-Shoresh ha-Avodah*. Jerusalem, Merkaz, 1940.

Azriel ben Menahem of Gerona. *Peirush ha-Aggadot le-Rabbi Azri'el*. Edited by Isaiah Tishby. 2nd ed. Jerusalem: Magnes Press, 1982.

Bahya ben Asher. *Bei'ur al ha-Torah*. Edited by Chaim D. Chavel. 3 vols. Jerusalem: Mossad Harav Kook, 1971–72.

Barrow, John D. *The Book of Universes: Exploring the Limits of the Cosmos*. New York: W. W. Norton, 2011.

———. *The Origin of the Universe*. New York: Basic Books, 1994.

———, and Joseph Silk. *The Left Hand of Creation: The Origin and Evolution of the Expanding Universe.* 2nd ed. New York: Oxford University Press, 1993.

Berger, Abraham. "The Messianic Self-Consciousness of Abraham Abulafia: A Tentative Evaluation." In *Essays on Jewish Life and Thought: Presented in Honor of Salo Wittmayer Baron*, edited by Joseph L. Blau, Arthur Hertzberg, Philip Friedman, and Isaac Mendelsohn, 55–61. New York: Columbia University Press, 1959.

Boneh Yerushalayim. Jerusalem: Defus ha-Ivri, 1926.

Borowitz, Eugene. *Renewing the Covenant: A Theology for the Postmodern Jew.* Philadelphia: Jewish Publication Society, 1991.

Boyarin, Daniel. *Carnal Israel: Reading Sex in Talmudic Culture.* Berkeley: University of California Press, 1993.

Buber, Martin. *I and Thou.* Translated by Walter Kaufmann. New York: Charles Scribner's Sons, 1970.

Cordovero, Moses. *Pardes Rimmonim.* Jerusalem: Mordekhai Etyah, 1962.

———. *Tomer Devorah.* Warsaw: Joel Levensohn, 1873.

Croswell, Ken. *The Alchemy of the Heavens.* New York: Anchor Books, 1995.

———. *The Universe at Midnight: Observations Illuminating the Cosmos.* New York: The Free Press, 2001.

Darling, David. *Deep Time: The Journey of a Single Subatomic Particle from the Moment of Creation to the Death of the Universe—and Beyond.* New York: Dell, 1989.

David ben Judah he-Hasid. *The Book of Mirrors: Sefer Mar'ot ha-Zove'ot.* Edited by Daniel Chanan Matt. Chico, Calif.: Scholars Press, 1982.

David ben Solomon ibn Abi Zimra. *She'elot u-Teshuvot ha-Radbaz.* New York: Otzar Hasefarim, 1966.

Davies, Paul. *The Mind of God: The Scientific Basis for a Rational World.* New York: Simon and Schuster, 1992.

Dawkins, Richard. *The Blind Watchmaker: Why the Evidence of Evolution Reveals a Universe without Design.* New York: W. W. Norton, 1986.

Dennett, Daniel C. *Consciousness Explained*. Boston: Little, Brown and Company, 1991.

Dever, William G. *Did God Have a Wife?: Archaeology and Folk Religion in Ancient Israel*. Grand Rapids, Mich.: William B. Eerdmans, 2005.

Dov Baer ben Abraham of Mezhirech. *Maggid Devarav le-Ya'aqov*. Edited by Rivka Schatz-Uffenheimer. Jerusalem: Magnes Press, 1976.

———. *Or ha-Emet*. Edited by Levi Yitzhak of Berdichev. Bnei Brak: Yahadut, 1967.

———. *Or Torah*. Jerusalem, 1956.

Drees, Willem B. *Beyond the Big Bang: Quantum Cosmologies and God*. La Salle, Ill.: Open Court, 1990.

Dyson, Freeman. *Infinite in All Directions*. New York: Harper and Row, 1989.

Eddington, Arthur S. *The Nature of the Physical World*. Cambridge, England: Cambridge University Press, 1928.

———. *New Pathways in Science*. Cambridge, England: Cambridge University Press, 1935.

Eisen, Arnold. "Covenant." In *Contemporary Jewish Religious Thought: Original Essays on Critical Concepts, Movements, and Beliefs*, edited by Arthur A. Cohen and Paul Mendes-Flohr, 107–12. New York: Charles Scribner's Sons, 1987.

Elior, Rachel. *The Paradoxical Ascent to God: The Kabbalistic Theosophy of Habad Hasidism*. Translated by Jeffrey M. Green. Albany: State University of New York Press, 1993.

Fackenheim, Emil L. *The Jewish Return into History: Reflections in the Age of Auschwitz and a New Jerusalem*. New York: Schocken, 1978.

Falk, Marcia. *The Book of Blessings: New Jewish Prayers for Daily Life, the Sabbath, and the New Moon Festival*. San Francisco: HarperSanFrancisco, 1996.

Feld, Edward. *The Spirit of Renewal: Crisis and Response in Jewish Life*. Woodstock, Vt.: Jewish Lights, 1991.

Ferris, Timothy. *Coming of Age in the Milky Way*. New York: Doubleday, 1989.

———. *The Whole Shebang: A State-of-the-Universe(s) Report*. New York: Simon and Schuster, 1997.

Fishbane, Michael. *Sacred Attunement: A Jewish Theology*. Chicago: University of Chicago Press, 2008.

Flusser, David. *Jewish Sources in Early Christianity*. New York: Adama Books, 1987.

Freeman, Kathleen. *Ancilla to the Pre-Socratic Philosophers*. Cambridge: Harvard University Press, 1966.

Friedman, Richard Elliott. *The Disappearance of God: A Divine Mystery*. Boston: Little, Brown and Company, 1995.

Fritzsch, Harald. *The Creation of Matter: The Universe from Beginning to End*. New York: Basic Books, 1984.

Frymer-Kensky, Tikva. *In the Wake of the Goddesses: Women, Culture and the Biblical Transformation of Pagan Myth*. New York: Ballantine Books, 1993.

Gillman, Neil. *Sacred Fragments: Recovering Theology for the Modern Jew*. Philadelphia: Jewish Publication Society, 1990.

Gliksman, Pinhas Zelig. *Der Kotsker Rebbe*. Pietrikow: Pahlman, 1938.

Green, Arthur. *Ehyeh: A Kabbalah for Tomorrow*. Woodstock, Vt.: Jewish Lights, 2003.

———. *Radical Judaism: Rethinking God & Tradition*. New Haven: Yale University Press, 2010.

———. *Seek My Face, Speak My Name: A Contemporary Jewish Theology*. Northvale, N.J.: Jason Aronson, 1992.

———. "*Shekhinah*, the Virgin Mary, and the Song of Songs: Reflections on a Kabbalistic Symbol in Its Historical Context." *AJS Review* 26 (2002): 1–52.

Greene, Brian. *The Elegant Universe: Superstrings, Hidden Dimensions, and the Quest for the Ultimate Theory*. New York: Vintage Books, 2000.

———. *The Fabric of the Cosmos: Space, Time, and the Texture of Reality*. New York: Alfred A. Knopf, 2004.

———. *The Hidden Reality: Parallel Universes and the Deep Laws of the Cosmos*. New York: Alfred A. Knopf, 2011.

Gribbin, John. *In Search of the Big Bang: Quantum Physics and Cosmology*. New York: Bantam Books, 1986.

———. *In the Beginning: The Birth of the Living Universe*. Boston: Little, Brown and Company, 1993.

Guth, Alan N. *The Inflationary Universe: The Quest for a New Theory of Cosmic Origins*. Reading, Mass.: Addison-Wesley, 1997.

Hawking, Stephen W. *A Brief History of Time: From the Big Bang to Black Holes*. 2nd ed. New York: Bantam Books, 1998.

———. "The Edge of Spacetime." *New Scientist* 103 (16 August 1984): 10–14.

———. "Quantum Cosmology." In *Three Hundred Years of Gravitation*, edited by S. W. Hawking and W. Israel, 631–51. Cambridge, England: Cambridge University Press, 1987.

———. *The Universe in a Nutshell*. New York: Bantam Books, 2001.

———, and Leonard Mlodinow. *The Grand Design*. New York: Bantam Books, 2010.

Horowitz, Shabbetai Sheftel. *Shefa Tal*. Lemberg, 1859.

Hoyle, Fred. "Continuous Creation." *The Listener* 41:1054 (April 7, 1949): 567–68.

———. *Home Is Where the Wind Blows: Chapters from a Cosmologist's Life*. Mill Valley, Calif.: University Science Books, 1994.

———. "Man's Place in the Expanding Universe." *The Listener* 43:1102 (March 9, 1950): 419–24.

———. *The Nature of the Universe*. New York: Harper and Brothers, 1950.

Hurwitz, Siegmund. "Psychological Aspects in Early Hasidic Literature." In *Timeless Documents of the Soul*, edited by James Hillman, 151–239. Evanston, Ill.: Northwestern University Press, 1968.

Idel, Moshe. "*Hitbodedut* as Concentration in Ecstatic Kabbalah." In *Jewish Spirituality: From the Bible through the Middle Ages,* edited by Arthur Green, 405–38. New York: Crossroad, 1986.

———. "Infinities of Torah in Kabbalah." In *Midrash and Literature,* edited by Geoffrey H. Hartman and Sanford Budick, 141–57. New Haven: Yale University Press, 1986.

———. *Kabbalah: New Perspectives*. New Haven: Yale University Press, 1988.

Impey, Chris. *How It Ends: From You to the Universe*. New York: W. W. Norton, 2010.

Isaac the Blind. "Commentary on Sefer Yetsirah." Appendix to Gershom Scholem, *Ha-Qabbalah be-Provans,* edited by Rivka Schatz. Jerusalem: Academon, 1970.

Issachar Baer of Zlotshov. *Mevasser Tsedeq*. Berditchev, 1817.

Jacob ben Sheshet. *Sefer Ha-Emunah ve-ha-Bittahon*. In *Kitvei Ramban,* edited by Chaim D. Chavel, 2:339–448. Jerusalem: Mossad Harav Kook, 1964.

———. *Sefer Meshiv Devarim Nekhohim*. Edited by Georges Vajda. Jerusalem: Israel Academy of Sciences and Humanities, 1968.

Jacob Joseph ben Zevi Hirsh ha-Kohen Katz of Polonnoye. *Ben Porat Yosef*. Lemberg: Y. Shtand, 1866.

———. *Toledot Ya'aqov Yosef*. Koretz: Zevi Hirsh ben Aryeh Lev Shemu'el, 1780.

———. *Tsofnat Pa'neah*. Koretz: Zevi Hirsh Margaliot, 1782.

John of the Cross. *Ascent of Mount Carmel*. Translated by E. Allison Peers. Garden City, N.Y.: Image Books, 1958.

Judah ben Barzillai. *Peirush Sefer Yetsirah*. Edited by S. J. Halberstam. Berlin: M'kize Nirdamim, 1885.

Judah Halevi, *Sefer ha-Kozari*. Translated and edited by Yehuda Even Shmuel. Tel Aviv: Dvir, 1972.

Justin Martyr. *Dialogue with Trypho*. In *The Ante-Nicene Fathers*, edited by Alexander Roberts and James Donaldson, 1:194–270. Grand Rapids, Mich.: Eerdmans, 1950.

Kafka, Franz. *Parables and Paradoxes*. Edited by Nahum N. Glatzer. New York: Schocken, 1961.

Keter Shem Tov. New York: Kehot, 1972.

Kirshner, Robert P. *The Extravagant Universe: Exploding Stars, Dark Energy, and the Accelerating Cosmos*. Princeton: Princeton University Press, 2002.

Koidonover, Zevi Hirsch. *Qav ha-Yashar*. Lemberg: Bak, 1858.

Kook, Abraham Isaac. *Orot*. Jerusalem: Mossad ha-Rav Kook, 1961.

———. *Orot ha-Qodesh*. 4 vols. Jerusalem: Mossad ha-Rav Kook, 1953–64.

Krauss, Lawrence M. *A Universe from Nothing: Why There Is Something Rather than Nothing*. New York: The Free Press, 2012.

Kushner, Lawrence. *God Was in This Place and I, i Did Not Know: Finding Self, Spirituality and Ultimate Meaning*. 2nd ed. Woodstock, Vt.: Jewish Lights, 2016.

———. *The River of Light: Jewish Mystical Awareness*. Woodstock, Vt.: Jewish Lights, 1990.

Lao Tzu. *Tao Te Ching: The Classic Book of Integrity and the Way*. Translated by Victor H. Mair. New York: Bantam, 1990.

Lavi, Shim'on. *Ketem Paz*. Djerba, Tunisia: Jacob Haddad, 1940. Reprint, 2 vols. Jerusalem: Ahavat Shalom, 1981.

Lemonick, Michael D. *Echo of the Big Bang*. Princeton: Princeton University Press, 2003.

Levi Yitzhak of Berdichev. *Qedushat Levi*. Jerusalem: Pe'er, 1972.

Levinas, Emmanuel. *Nine Talmudic Readings*. Translated by Annette Aronowicz. Bloomington, Ind.: Indiana University Press, 1990.

Levine, Amy-Jill. *The Misunderstood Jew: The Church and the Scandal of the Jewish Jesus*. San Francisco, Calif.: HarperSanFrancisco, 2006.

Lightman, Alan, and Roberta Brawer, *Origins: The Lives and Worlds of Modern Cosmologists*. Cambridge: Harvard University Press, 1990.

Linde, Andrei. "Inflation and Quantum Cosmology." In *Three Hundred Years of Gravitation*, edited by S.W. Hawking and W. Israel, 604–30. Cambridge, England: Cambridge University Press, 1987.

———. "Particle Physics and Inflationary Cosmology." *Physics Today* 40:9 (1987): 61–68.

———. "The Self–Reproducing Inflationary Universe." *Scientific American* (November 1994): 48–55.

Lipman, Eliezer. *Tal Orot im Migdal David*. Vienna: Yozef Hrashantski, 1792.

Liqqutei Yeqarim. Lemberg, 1865.

Liqqutim Hadashim me-ha-Ari u-mi-Maharhu. Jerusalem: Mevaqqeshei ha-Shem, 1985.

Ma'arekhet ha-Elohut. Mantua, Meir ben Efraim, 1558. Reprint, Jerusalem: Meqor Hayyim, 1963.

Maimonides, Moses. *The Guide of the Perplexed*. Translated by Shlomo Pines. Chicago: University of Chicago Press, 1963.

Marcus Aurelius. *Meditations*. Translated by Maxwell Staniforth. Harmondsworth, England: Penguin, 1964.

Matt, Daniel C. "*Ayin*: The Concept of Nothingness in Jewish Mysticism." In *Essential Papers on Kabbalah*, edited by Lawrence Fine, 67–108. New York: New York University Press, 1995.

———. *The Essential Kabbalah: The Heart of Jewish Mysticism*. San Francisco, Calif.: HarperSanFrancisco, 1995.

———. "The Mystic and the *Mizwot*." In *Jewish Spirituality: From the Bible through the Middle Ages*, edited by Arthur Green, 367–404. New York: Crossroad, 1986.

———. "Varieties of Mystical Nothingness: Jewish, Christian and Buddhist." *The Studia Philonica Annual: Studies in Hellenistic Judaism* 9 (1997): 316–31.

———, trans. and ed. *The Zohar: Pritzker Edition.* Vols. 1–9. Stanford, Calif.: Stanford University Press, 2004–16.

———, trans. and ed. *Zohar: The Book of Enlightenment.* Mahwah, N.J.: Paulist Press, 1983.

McGinn, Bernard. "The God beyond God: Theology and Mysticism in the Thought of Meister Eckhart." *The Journal of Religion* 61 (1981): 1–19.

Menahem Nahum of Chernobyl. *Me'or Einayim.* New York: Twersky Brothers, 1952.

Michaelson, Jay. *Everything Is God: The Radical Path of Nondual Judaism.* Boston: Trumpeter, 2009.

Midrash Tanhuma ha-Qadum ve-ha-Yashan. Edited by Solomon Buber. 3 vols. Vilna: Romm, 1885.

Misner, Charles W., Kip S. Thorne, John Archibald Wheeler. *Gravitation.* San Francisco, W. H. Freeman, 1973.

Mitton, Simon. *Fred Hoyle: A Life in Science.* Cambridge, England: Cambridge University Press, 2011.

Moltmann, Jürgen. *God in Creation: A New Theology of Creation and the Spirit of God.* San Francisco: Harper and Row, 1985.

Moses ben Shem Tov de León. *The Book of the Pomegranate: Moses de León's Sefer ha-Rimmon.* Edited by Elliot R. Wolfson. Atlanta: Scholars Press, 1988.

———. *Sheqel ha-Qodesh.* Edited by Charles Mopsik. Los Angeles: Cherub Press, 1996.

Nelson, David W. *Judaism, Physics and God: Searching for Sacred Metaphors in a Post-Einstein World.* Woodstock, Vt.: Jewish Lights, 2005.

Ornstein, Robert. *The Evolution of Consciousness.* New York: Prentice Hall, 1991.

Osserman, Robert. *Poetry of the Unvierse: A Mathematical Exploration of the Cosmos.* New York: Doubleday, 1995.

Overbye, Dennis. *Lonely Hearts of the Cosmos: The Story of the Scientific Quest for the Secret of the Universe.* New York: HarperCollins, 1991.

Pagels, Heinz. *Perfect Symmetry: The Search for the Beginning of Time.* New York: Simon and Schuster, 1985.

Pais, Abraham. *"Subtle Is the Lord ...": The Science and the Life of Albert Einstein.* Oxford: Oxford University Press, 1982.

Panek, Richard. *The 4 Percent Universe: Dark Matter, Dark Energy, and the Race to Discover the Rest of Reality.* Boston: Houghton Mifflin Harcourt, 2011.

Patai, Raphael. *The Hebrew Goddess.* 3rd ed. Detroit: Wayne State University Press, 1990.

Primack, Joel R., and Nancy E. Abrams, "'In a Beginning ...': Quantum Cosmology and Kabbalah." *Tikkun* 10:1 (January–February 1995): 66–73.

———, and Nancy Ellen Abrams. *The New Universe and the Human Future: How a Shared Cosmology Could Transform the World.* New Haven: Yale University Press, 2011.

———, and Nancy Ellen Abrams. *The View from the Center of the Universe: Discovering Our Extraordinary Place in the Cosmos.* New York: Riverhead Books, 2006.

Rees, Martin. *Before the Beginning: Our Universe and Others.* Reading, Mass,: Addision Wesley, 1997.

———. *Just Six Numbers: The Deep Forces That Shape the Universe.* New York: Basic Books, 1999.

Reeves, Hubert. "Birth of the Myth of the Birth of the Universe." In *New Windows to the Universe,* Volume 2, edited by F. Sanchez and M. Vasquez, 141–49. Cambridge, England: Cambridge University Press, 1990.

Roos, Matt. *Introduction to Cosmology.* 4th ed. West Sussex, England: John Wiley & Sons, 2015.

Rubenstein, Richard L. *After Auschwitz: History, Theology, and Contemporary Judaism.* 2nd ed. Baltimore: Johns Hopkins University Press, 1992.

Sagan, Carl, and Ann Druyan. *Shadows of Forgotten Ancestors: A Search for Who We Are.* New York: Random House, 1992.

Samuelson, Norbert M. *Judaism and the Doctrine of Creation*. Cambridge, England: Cambridge University Press, 1994.

Sanders, E. P. *Jesus and Judaism*. Philadelphia: Fortress Press, 1985.

———. "The Life of Jesus." In *Christianity and Judaism: A Parallel History of Their Origins and Development*, edited by Hershel Shanks, 41–83. Washington, D.C.: Biblical Archaeology Society, 1992.

Sarug, Israel. *Limmudei Atsilut*. Munkacs: Blayer and Kahn, 1897.

Schimmel, Annemarie. *Mystical Dimensions of Islam*. Chapel Hill: University of North Carolina Press, 1975.

Scholem, Gershom. *Devarim be-Go*. 2 vols. Tel Aviv: Am Oved, 1976.

———. *Kabbalah*. Jerusalem: Keter, 1974.

———. *Major Trends in Jewish Mysticism*. New York: Schocken, 1961.

———. *The Messianic Idea in Judaism*. New York: Schocken, 1971.

———. *On the Kabbalah and Its Symbolism*. Translated by Ralph Manheim. New York: Schocken, 1969.

———. *On the Mystical Shape of the Godhead: Basic Concepts in Kabbalah*. Translated by Joachim Neugroschel, edited by Jonathan Chipman. New York: Schocken, 1991.

———. *Sabbatai Sevi: The Mystical Messiah*. Princeton, N.J.: Princeton University Press, 1973.

———. *Über einige Grundbegriffe des Judentums*. Frankfurt: Suhrkamp Verlag, 1970.

Schroeder, Gerald L. *Genesis and the Big Bang: The Discovery of Harmony Between Modern Science and the Bible*. New York: Bantam, 1990.

Searle, John. *The Rediscovery of the Mind*. Cambridge, Mass.: MIT Press, 1992.

Shivhei ha-Besht. Edited by Benjamin Mintz. Tel Aviv: Talpiyot, 1961.

Shneur Zalman of Lyady. *Torah Or*. Vilna: Romm, 1899.

Silk, Joseph. *On the Shores of the Unknown: A Short History of the Universe*. Cambridge, England: Cambridge University Press, 2005.

Singh, Simon. *Big Bang: The Origin of the Universe*. New York: HarperCollins, 2004.

Smith, Howard. *Let There Be Light: Modern Cosmology and Kabbalah; A New Conversation between Science and Religion*. Novato, Calif.: New World Library, 2006.

Smoot, George, and Keay Davidson. *Wrinkles in Time*. New York: Avon, 1994.

Swimme, Brian, and Thomas Berry. *The Universe Story: From the Primordial Flaring Forth to the Ecozoic Era*. San Francisco: HarperSanFrancisco, 1992.

Thorne, Kip S. *Black Holes and Time Warps: Einstein's Outrageous Legacy*. New York: W. W. Norton, 1994.

Tyson, Neil deGrasse, and Donald Goldsmith. *Origins: Fourteen Billion Years of Cosmic Evolution*. New York: W. W. Norton, 2004.

The Upanishads. Translated by Swami Prabhavananda and Frederick Manchester. New York: Mentor Books, 1957.

Vermes, Geza. *Jesus the Jew: A Historian's Reading of the Gospels*. Philadelphia: Fortress Press, 1973.

Vidas, Elijah ben Moses de. *Reshit Hokhmah*. Amsterdam: Nathanel Foa, 1708.

Weinberg, Steven. *Dreams of a Final Theory: The Scientist's Search for the Ultimate Laws of Nature*. New York: Vintage Books, 1994.

———. *The First Three Minutes: A Modern View of the Origin of the Universe*. 2nd ed. New York: Basic Books, 1993.

Wheeler, John Archibald. *Geons, Black Holes, and Quantum Foam: A Life in Physics*. With Kenneth Ford. New York: W. W. Norton, 1998.

———. *A Journey into Gravity and Spacetime*. New York: Scientific American Library, 1990.

Wilczek, Frank. *A Beautiful Question: Finding Nature's Deep Design*. New York: Penguin Press, 2015.

Wolpe, David J. *In Speech and in Silence: The Jewish Quest for God*. New York: Henry Holt and Company, 1992.

Wyschogrod, Michael. *The Body of Faith: Judaism as Corporeal Election*. New York: Seabury Press, 1983.

Zajonc, Arthur. *Catching the Light: The Entwined History of Light and Mind*. New York: Oxford University Press, 1993.

Zee, A. *An Old Man's Toy: Gravity at Work and Play in Einstein's Universe*. New York: Collier Books, 1989.

Zev Wolf of Zhitomir. *Or ha-Me'ir*. Warsaw: Meir Halter, 1883.

Printed in the USA
CPSIA information can be obtained
at www.ICGtesting.com
JSHW022332140824
68134JS00019B/1444